CHEMICAL AND MICROBIOLOGICAL ANALYSIS OF MILK PRODUCTS

CHEMICAL AND MICROBIOLOGICAL ANALYSIS OF MILK PRODUCTS

Dr. Veena Kumari

RANDOM PUBLICATIONS
NEW DELHI - 110 002 (INDIA)

Chemical and Microbiological Analysis of Milk Products

ISBN 978-93-51116-97-4

Published in 2015 in India by

RANDOM PUBLICATIONS

4376-A/4B, Gali Murari Lal, Ansari Road
New Delhi-110 002
Phone: +9111-43580356, 23289044
E-mail: randomexports@gmail.com; sales@randompublications.com; info@randompublications.com

Reprinted 2019

Type Setting by: Friends Media, Delhi-110089
Digitally Printed at: Replika Press Pvt. Ltd.

Preface

Each type of food has its own structure, chemical composition, moisture content, acidity, and oxygen tension. These factors decide the predominant microflora of the normal product, the vulnerability to certain microorganisms, and the potential danger of the food to humans. Milk is a white liquid produced by the mammary glands of mammals. It is the primary source of nutrition for young mammals before they are able to digest other types of food. Early-lactation milk contains colostrum, which carries the mother's antibodies to the baby and can reduce the risk of many diseases in the baby. It also contains many other nutrients. In many cultures of the world, especially the Western world, humans continue to consume milk beyond infancy, using the milk of other animals (especially cattle, goats and sheep) as a food product. Initially, the ability to digest milk was limited to children as adults did not produce lactase, an enzyme necessary for digesting the lactose in milk. Milk was therefore converted to curd, cheese and other products to reduce the levels of lactose. Thousands of years ago, a chance mutation spread in human populations in Europe that enabled the production of lactase in adulthood. This allowed milk to be used as a new source of nutrition which could sustain populations when other food sources failed. Milk is processed into a variety of dairy products such as cream, butter, yogurt, kefir, ice cream, and cheese. Modern industrial processes use milk to produce casein, whey protein, lactose, condensed milk, powdered milk, and many other food-additive and industrial products. Milk is an emulsion or colloid of butterfat globules within a water-based fluid that contains dissolved carbohydrates and protein aggregates with minerals. Normal bovine milk contains 30–35 grams of protein per litre of which about 80% is arranged in casein micelles. The largest structures in the fluid portion of the milk are "casein micelles": aggregates of several thousand protein molecules with superficial resemblance to a surfactant micelle, bonded with the help of nanometer-scale particles of calcium phosphate. Each casein micelle is roughly spherical and about a tenth of a micrometer across. There are four different types of casein proteins: αs1-, αs2-, β-, and κ-caseins. Collectively, they make up around 76–86% of the protein in milk, by weight.

Minerals or milk salts, are traditional names for a variety of cations and anions within bovine milk. Calcium, phosphate, magnesium, sodium, potassium, citrate, and chlorine are all included as minerals and they typically occur at concentration of 5–40 mM. The milk salts strongly interact with casein, most notably calcium phosphate. It is present in excess and often, much greater excess of solubility of solid calcium phosphate. In addition to calcium, milk is a good source of many other vitamins. Vitamins A, B6, B12, C, D, K, E, thiamine, niacin, biotin, riboflavin, folates, and pantothenic acid are all present in milk. Milk contains several different carbohydrate including lactose, glucose, galactose, and other oligosaccharides. The lactose gives milk its sweet taste and contributes approximately 40% of whole cow's milk's calories. Lactose is a disaccharide composite of two simple sugars, glucose and galactose. Bovine milk averages 4.8% anhydrous lactose, which amounts to about 50% of the total solids of skimmed milk. Levels of lactose are dependent upon the type of milk as other carbohydrates can be present at higher concentrations that lactose in milks. Other components found in raw cow's milk are living white blood cells, mammary gland cells, various bacteria, and a large number of active enzymes. When raw milk is left standing for a while, it turns "sour". This is the result of fermentation, where lactic acid bacteria ferment the lactose in the milk into lactic acid.

This valuable resource book will satisfy the need of students, teachers, and all those concerned with dairy industry and utilisation of their products.

I thank all members of my team who have helped in the preparation of the book. My special thanks go to "Random Publications" who have published the book.

— Dr. Veena Kumari

Contents

Chapter 1

Introduction

Food spoilage is an enormous economic problem worldwide. Through microbial activity alone, approximately one-fourth of the world's food supply is lost. Milk is a highly nutritious food that serves as an excellent growth medium for a wide range of microorganisms. The microbiological quality of milk and dairy products is influenced by the initial flora of raw milk, the processing conditions, and post-heat treatment contamination. Undesirable microbes that can cause spoilage of dairy products include Gram-negative psychrotrophs, coliforms, lactic acid bacteria, yeasts, and molds.

In addition, various bacteria of public health concern such as Salmonella spp., Listeria monocytogenes, Campylobacter jejuni, Yersinia enterocolitica, pathogenic strains of Escherichia coli and enterotoxigenic strains of Staphylococcus aureus may also be found in milk and dairy products. For this reason, increased emphasis should be placed on the microbiological examination of milk and dairy foods.

Microbiological analyses are critical for the assessment of quality and safety, conformation with standards and specifications, and regulatory compliance.

In Hungary, a very comprehensive set of microbiological reference criteria covering virtually all foods and including limits for an extensive range of indicators and pathogens was issued by the Ministry of Health in 1998. This official document entitled "Decree No. 4/1998 (XI. 11.) EüM on the acceptable levels of microbiological contamination in foods" is based on ICMSF and FAO principles in that it includes sampling plans and identifies the criteria for acceptance or rejection of a lot. Altogether 26 food groups have mandatory criteria and the number or type of microorganisms for which there are limits for any particular food varies from one to nine.

Divided into nine subgroups, milk and milk products constitute one of the 26 groups formed. As of 2003, in terms of dairy foods, this legal document was replaced by another one titled "Decree No. 1/2003 (I. 8.) FVM-ESzCsM on the food hygienic conditions for production and placing on the market of raw milk, heat-treated milk and milk-based products", which is identical in content with the EU Milk Hygiene Directive 92/46/EEC.

Production and use of Microbial Enzymes for Dairy Processing

Indians are known to be lovers of milk and its products. As a sequel to white revolution, India has surged ahead to become the largest milk producer in the world, the production figure touched to 74 million tons for the year 1998. With growing urbanization, demand for processed dairy foods has increased considerably, in particular demand for different cheese varieties, and low-lactose milk due to increasing intolerance of human beings to lactose in milk and other milk products. For improving the quality of milk and milk products, a number of different enzymes from microbial as well as from nonmicrobial sources have potential applications in dairy processing.

The use of rennet in cheese manufacture was among the earliest applications of exogenous enzymes in food processing, dating back to approximately 6000 B C. In 1994, the total production of cheese was 8000 metric tons against a total demand of 9000 metric tons. The projected demand by 2000 A D is around 30,000 metric tons. The use of rennet, as an exogenous enzyme, in cheese manufacture is perhaps the largest single application of enzymes in food processing.

In recent years, proteinases have found additional applications in dairy technology, for example in acceleration of cheese ripening, modification of functional properties and preparation of dietic products. Principal among some enzymes that have important and growing applications are lipases and b-galactosidases. Enzymes with limited applications include glucose oxidase, superoxide dismutase, sulphydryl oxidase, etc. The increasing use of enzymes to produce specific products with characteristic attributes can be emphasized by the world-wide sale of industrial enzymes approximating to US $ 1.6 billion, which is expected to reach US $ 3.0 billion by the year 2008. Almost 45% of this is shared by the food industry and the remaining is shared by detergent (34.4%), textile (11%), leather (2.8%), pulp and paper (1.2%) industries and other industries (5.6%) excluding enzymes for use in diagnostics and therapeutics. India being the highest producer of milk in the world, and consequently the surplus availability of milk in our country has triggered the food and dairy industry to convert the liquid milk into value-added products using biochemical and enzymatic processes.

Microbial Rennets in Dairy Applications

Animal rennet (bovine chymosin) is conventionally used as a milk-clotting agent in dairy industry for the manufacture of quality cheeses with good flavour and texture. Owing to an increase in demand for cheese production world wide – i.e. 4% per annum over the past 20 years, approximating 13.533 million tons – coupled with reduced supply of calf rennet, has therefore led to a search for rennet substitutes, such as microbial rennets. At present, microbial rennet is used for one-third of all the cheese produced world wide. Rennin acts on the milk protein in two stages, by enzymatic and by nonenzymatic action, resulting in coagulation of milk. In the enzymatic phase, the resultant milk becomes a gel due to the influence of calcium ions and the temperature used in the process. Many microorganisms are known to produce rennet-like proteinases which can substitute the calf rennet.

Microorganisms like Rhizomucor pusillus, R. miehei, Endothia parasitica, Aspergillus oryzae, and Irpex lactis are used extensively for rennet production in cheese manufacture. Extensive research that has been carried out so far on rennet substitutes has been reviewed by several authors. Different strains of species of Mucor are often used for the production of microbial rennets. Whereas best yields of the milk-clotting protease from Rhizomucor pusillus are obtained from semisolid cultures containing 50% wheat bran, R. miehei and Endothia parasitica are well suited for submerged cultivation. Using the former, good yields of milk-clotting protease may be obtained in a medium containing 4% potato starch, 3% soybean meal, and 10% barley. During growth, lipase is secreted together with the protease. Therefore, the lipase activity has to be destroyed by reducing the pH, before the preparation can be used as cheese rennet.

Microbial rennets from various microorganisms (marketed under the trade names such as Rennilase, Fromase, Marzyme, Hanilase, etc.) being marketed since the 1970s have proved satisfactory for the production of different kinds of cheese. The molecular and enzymatic properties of chymosins have been studied extensively. Although the proteolytic specificities of the three commonly used fungal rennets are considerably different from those of calf chymosin, these rennets have been used to produce acceptable cheeses.

Recently Novo Nordisk has succeeded in expressing just one proteolytic enzyme from the fungus R. miehei in the well-known organism A. oryzae. This host organism is able to produce the single protease that cleaves the ·-casein into a glycomacropeptide and para casein by hydrolysing only at the phe105–met106 peptide bond between phenyl alanine and methionine.

This monocomponent enzyme product has the trade name Novoren. One major drawback of microbial rennet use in cheese manufacture, is the development of off flavour and bitter taste in the nonripened as well as in the ripened cheeses. The rennets from microbial sources are more proteolytic in nature in comparison to rennet from animal sources, resulting in production of some bitter peptides during the process of cheese ripening. Hence, attempts have been made to clone the gene for calf chymosin, and to express it in selected bacteria, yeasts, and molds.

Recombinant Rennets for Cheese Manufacture

Due to shortage of calf stomachs and the economic value of cheese rennet, gene for calf chymosin was one of the first genes for mammalian enzymes that was cloned and expressed in microorganisms. Many different laboratories have cloned the gene for calf prochymosin in Escherichia coli, and analysed the structure of the gene as well as the properties of the recombinant chymosin. The expressed proenzyme in E. coli, is present mainly as insoluble inclusion bodies, comprised of reduced prochymosin as well as molecules that are interlinked by disulphide bridges. After disintegration of the cells, inclusion bodies are harvested by centrifugation.

The individual laboratories have reported some differences in the procedure for renaturation of prochymosin from the inclusion bodies, but all have followed the same general scheme. The enzymatic properties of recombinant E. coli chymosin are indistinguishable from those of native calf chymosin. The enzymes were identical when observed by immunodiffusion in gels, but a slight difference was observed by enzyme linked immunosorbent assay (ELISA). The gene for prochymosin has also been cloned in Saccharomyces cerevisiae; the levels of expression have been reported to be 0.5 to 2.0% of total yeast protein. In yeast, about 20% of the prochymosin can be released in soluble form which can be activated directly; the remaining 80% is still associated with the cell debris.

The cloning of chymosin was carried out without the prosequence. Though the level of chymosin mRNA was similar to that of prochymosin mRNA, no milk-clotting activity was observed in clones containing the chymosin gene only. The results suggest that the prosequence is essential for correct folding of the polypeptide chain. The zymogen for the aspartic proteases from R. pusillus, also called mucor rennin, has likewise been cloned and expressed in yeast. Studies on its conversion to active form showed that secretion of R. pusillus protease from recombinant yeast was dependent on glycosylation of the enzyme.

Compared to yeast, the filamentous fungi generally secrete larger quantities of proteins into the culture medium. Furthermore, filamentous fungi secrete the heterologous proteins with correct folding of the polypeptide chain, and process correct pairing of sulphydryl groups. The gene for R. miehei protease has been expressed in A. oryzae. Prochymosin has been expressed in Kluyveromyces lactis, A. nidulans, A. niger, and Trichoderma reesei. In most of the cases, the reported yields of the model systems were about 10–40 mg of enzyme per litre of culture medium.

However, 3.3 g of enzyme per liter has also been achieved. Yeast, Kluyveromyces lactis, has recently been used as an efficient host for the secretion of recombinant chymosin, which has led to the development of large-scale production process for chymosin. If produced on an industrial scale, the yields will perhaps be of the latter magnitude. Most of the companies produce recombinant rennet of cattle calf origin in different microbial hosts, however in India, the major source of milk is buffalo, showing a different composition from that of cow. The rennet from buffalo source has inherent compatibility for clotting buffalo milk. Therefore, National Dairy Research Institute, Karnal, has taken a lead in cloning gene of buffalo chymosin. The buffalo chymosin cDNA has been cloned in E. coli and partial N-terminal sequencing of the purified buffalo chymosin indicates that it is highly homologous to the cattle chymosin.

Several cheese making experiments have been carried out with recombinant chymosin and the general aspects of recombinant chymosin have been dealt with in a report. Since most of the rennet (> 90 %) added to cheese milk is lost in the whey, immobilization would considerably extend its catalytic life. Several rennets have been immobilized, but their efficiency as milk coagulants has been questioned. So, there is a fairly general support for the view that immobilized enzymes cannot coagulate milk properly owing to inaccessability of the Phe–Met peptide bond of k-casein, and that the apparent coagulating activity of immobilized rennets is due to leaching of the enzyme from the support.

Different types of conventional cheeses have been successfully made by using recombinant rennet on an experimental or pilot scale. No major differences have been detected between cheeses made with recombinant chymosins or natural enzymes, regarding cheese yield, texture, smell, taste and ripening. The recombinant chymosins are identical with calf rennet according to the report on biochemical and genetic evidences.

Lactase in Dairy Industry

Lactose, the sugar found in milk and whey, and its corresponding hydrolase, lactase or b-galactosidase, have been extensively researched during the past decade. This is because of the enzyme immobilization technique which has given new and interesting possibilities for the utilization of this sugar. Because of intestinal enzyme insufficiency, some individuals, and even a population, show lactose intolerance and difficulty in consuming milk and dairy products.

Hence, low-lactose or lactose-free food aid programme is essential for lactose-intolerant people to prevent severe tissue dehydration, diarrhoea, and, at times, even death. Another advantage of lactase-treated milk is the increased sweetness of the resultant milk, thereby avoiding the requirement for addition of sugars in the manufacture of flavoured milk drinks. Manufacturers of ice cream, yoghurt and frozen desserts use lactase to improve scoop and creaminess, sweetness, and digestibility, and to reduce sandiness due to crystallization of lactose in concentrated preparations. Cheese manufactured from hydrolysed milk ripens more quickly than the cheese manufactured from normal milk.

Technologically, lactose crystallizes easily which sets limits to certain processes in the dairy industry, and the use of lactase to overcome this problem has not reached its fullest potential because of the associated high costs. Moreover, the main problem associated with discharging large quantities of cheese whey is that it pollutes the environment. But, the discharged whey could be exploited as an alternate cheap source of lactose for the production of lactic acid by fermentation. The whey permeate, which is a by-product in the manufacture of whey protein concentrates, by ultrafiltration could be fermented efficiently by Lactobacillus bulgaricus.

Lactose can be obtained from various sources like plants, animal organs, bacteria, yeasts (intracellular enzyme), or molds. Some of these sources are used for commercial enzyme preparations. Lactase preparations from A. niger, A. oryzae, and Kluyveromyces lactis are considered safe because these sources already have a history of safe use and have been subjected to numerous safety tests. The most investigated E. coli lactase is not used in food processing because of its cost and toxicity problems.

Properties of Lactase

The properties of the enzyme depend on its source. Temperature and pH optima differ from source to source and with the type of particular commercial preparation. Immobilization of the enzymes, method of immobilization, and type of carrier can also influence these optima values. In general, fungal lactase have pH optima in the acidic range 2.5–4.5, and yeast and bacterial lactases in the neutral region 6–7 and 6.5–7.5, respectively.

The variation in pH optima of lactases makes them suitable for specific applications, for example fungal lactases are used for acid whey hydrolysis, while yeast and bacterial lactases are suitable for milk (pH 6.6) and sweet whey (pH 6.1) hydrolysis. Product inhibition, e.g. inhibition by galactose, is another property which also depends on the source of lactase. The enzyme from A. niger is more strongly inhibited by galactose than that from A. oryzae.

This inhibition can be overcome by hydrolysing lactose at low concentrations by using immobilized enzyme systems or by recovering the enzyme using ultrafiltration after batch hydrolysis. Lactase from Bacillus species are superior with respect to thermostability, pH operation range, product inhibition, and sensitivity against high-substrate concentration. Thermostable enzymes, able to retain their activity at 60°C or above for prolonged periods, have two distinct advantages viz. they give higher conversion rate or shorter residence time for a given conversion rate, and the process is less prone to microbial contamination due to higher operating temperature. Bacillus species have a pH optima of 6.8 and a temperature optima of 65°C. Its high activity for skim milk, and less inhibition by galactose has made it suitable for use as a production organism for lactase.

The enzymatic hydrolysis of lactose can be achieved either by free enzymes, usually in batch fermentation process, or by immobilized enzymes or even by immobilized whole cells producing intracellular enzyme.

Although numerous hydrolysis systems have been investigated, only few of them have been scaled up with success and even fewer have been applied at an industrial or semi-industrial level. Several acid hydrolysis systems have been developed to industry-scale level. Large-scale systems which use free enzyme process have been developed for processing of UHT-milk and processing of whey, using K. lactis lactase (Maxilact, Lactozyme). Several commercial immobilized systems have been developed for commercial exploitation. Snamprogretti process of industrial-scale milk processing technology in Italy is one such working systems.

They make use of fibre-entrapped yeast lactase in a batch process, and the milk used is previously sterilized by UHT. For pilot plants, there are three other processes designed and developed to handle milk; (i) by Gist-Brocades, Rohm GmbH (Germany), and (ii) by Sumitomo, Japan. These are continuous processes with short residence times. Processing of whey UF-permeate is accomplished by the system developed by Corning Glass, Connecticut, Lehigh, Valio and Amerace corp. The process by Corning Glass is being applied at commercial scale in the bakers yeast production using hydrolysed whey.

Microbial Enzymes in Accelerated Cheese Ripening

Cheese ripening is a complex process mediated by biochemical and biophysical changes during which a bland curd is developed into a mature cheese with characteristic flavour, texture, and aroma. The desirable attributes are produced by the partial and gradual breakdown of carbohydrates, lipids, and proteins during ripening, mediated by several agents, viz. (i) residual coagulants, (ii) starter bacteria and their enzymes, (iii) nonstarter bacteria and their enzymes, (iv) indegenous milk enzymes, especially proteinases, and (v) secondary inocula with their enzymes.

Proteolysis occurs in all the cheese varieties and is a prerequisite for characteristic flavour development that can be regulated by proper use of the above agents. Cheese ripening is essentially an enzymatic process which can be accelerated by augmenting activity of the key enzymes. This has the advantage of initiating more specific action for flavour development compared to use of elevated temperatures that can result in accelerating undesirable nonspecific reactions, and consequently off flavour development. Enzymes may be added to develop specific flavours in cheeses, for example lipase addition for the development of Parmesan or Blue-type cheese flavours. Attempts to accelerate the multiple secondary flavour-forming-reactions, e.g. Strecker degradation, have been scarce. The pathways leading to the formation of flavour compounds are largely unknown, and therefore the use of exogenous enzymes to accelerate ripening is mostly an empirical process. Studies on the hydrolysis of whole casein and isolated casein components to observe the kinetics and specificity of aspartic proteases in rennin, pepsin, and four microbial rennet substitutes indicated the considerable differences in the reaction velocity and the extent of hydrolysis on the rennet curd yield. The rennet enzymes were active even in the later phases of cheese-ripening.

Proteinases and Peptidases

Proteolysis is characteristic of most cheese varieties and is indispensable for good flavour and textural development. Proteinases used in cheese processing include (i) plasmin, (ii) rennet, and (iii) proteinases (cell wall and/or intracellular) of the starter and nonstarter bacteria. Approximately 6% of the rennet added to cheese milk remains in the curd after manufacture and contributes significantly to proteolysis during ripening.

Combinations of individual neutral proteinases and microbial peptidases intensified cheese flavour, and when used in combinations with microbial rennets reduced the intensity of bitterness caused by the latter. Acid proteases in isolation cause intense bitterness. Various animal or microbial lipases gave pronounced cheese flavour, low bitterness and strong rancidity, while lipases in combination with proteinases and/or peptidases give good cheese flavour with low levels of bitterness. In a more balanced approach to the acceleration of cheese ripening using mixtures of proteinases and peptidases, attenuated starter cells or cell-free extracts (CFE) are being favoured.

Proteolytic Enzymes of Lactic Acid Bacteria in Fermented Milk Products

The proteolytic system of lactic acid bacteria is essential for their growth in milk, and contributes significantly to flavour development in fermented milk products. The proteolytic system is composed of proteinases which initially cleaves the milk protein to peptides; peptidases which cleave the peptides to small peptides and amino acids; and transport system responsible for cellular uptake of small peptides and amino acids.

Lactic acid bacteria have a complex proteolytic system capable of converting milk casein to the free amino acids and peptides necessary for their growth. These proteinases include extracellular proteinases, endopeptidases, aminopeptidases, tripeptidases, and proline-specific peptidases, which are all serine proteases. Apart from lactic streptococcal proteinases, several other proteinases from nonlactostreptococcal origin have been reported. There are also serine type of proteinases, e.g. proteinases from Lactobacillus acidophillus, L. plantarum, L. delbrueckii sp. bulgaricus, L. lactis, and L. helveticus. Aminopeptidases are important for the development of flavour in fermented milk products, since they are capable of releasing single amino acid residues from oligopeptides formed by extracellular proteinase activity.

Other Dairy Enzymes

Other enzymes used for dairy food application include: (i) proteases to reduce allergic properties of cow milk products for infants, and (ii) lipases for development of lipolytic flavours in speciality cheeses. The functional properties of milk proteins may be improved by limited proteolysis through the enzymatic modification of milk proteins. An acid-soluble casein, free of off flavour and suitable for incorporation into beverages and other acid foods, has been prepared by limited proteolysis. The antigenicity of casein is destroyed by proteolysis, and the hydrolysate is suitable for use in milk-protein-based foods for infants allergic to cow milk.

Lipolysis makes an important contribution to swiss cheese flavours, due mainly to the lipolytic enzymes of the starter cultures. The characteristic peppery flavour of Blue cheese is due to short-chain fatty acids and methyl ketones. Most of the lipolysis in Blue cheese is catalysed by Penicillium roqueforti lipase, with a lesser contribution from indigenous milk lipase. The NOVO process for production of enzyme modified cheese (EMC) uses medium-aged cheese which is emulsified, homogenized, and pasteurized, after which 'palatase' (a lipase from R. miehei) is added, with or without a proteinase, and the blend is ripened at a high temperature for one to four days.

The mixture is reheated, a paste results which is suitable for inclusion in soups, dips, dressings, or snack foods. EMC technology has been developed to produce a range of characteristic cheese flavours and flavour intensities, for example swiss, blue, cheddar, provolo-nemor or romano, suitable for inclusion at low levels in many products. The claims that exogenous enzymes are effective in accelerating ripening have not led to their wide-spread use, possibly due to their high cost, difficulties in distributing them uniformly in the curd, and the possible danger of over-ripening the cheese.

The other minor enzymes having limited applications in dairy processing include glucose oxidase, catalase, superoxide dismutase, sulphydryl oxidase, lactoperoxidase, and lysozymes. Glucose oxidase and catalase are often used together in selected foods for preservation. Superoxide dismutase is an antioxidant for foods and generates H_2O_2, but is more effective when catalase is present.

Thermally induced generation of volatile sulphydryl groups is thought to be responsible for the cooked off-flavour in ultra high temperature (UHT) processed milk. Use of sulphydryl oxidase under aseptic conditions can eliminate this defect. The natural inhibitory mechanism in raw milk is due to the presence of low levels of lactoperoxidase (LP), which can be activated by the external addition of traces of H_2O_2 and thiocyanate. It has been reported that the potential of LP-system and its activation enhances the keeping quality of milk.

Cow milk can be provided with protective factors by the addition of lysozyme, making it suitable as an infant milk. Lysozyme acts as a preservative by reducing bacterial counts in milk without affecting the L. bifidus *activity. The scope of application of minor enzymes to milk and milk products has been recently reviewed.*

The global market for the production of microbial enzymes for use in dairy-products manufacture is considerably large, but is being dominated only by a limited number of enzyme producers. In India the microbial dairy enzymes requirement has been very limited till now. However, with the advent of technological processes for the manufacture of different varieties of milk products, such as cheeses by the State Dairy Federations, Co-operatives and Private Dairy Product Manufacturers like Amul, Vijaya, Verka, Dynamix, Nestle, Smith Kline Beechem, etc., the markets for the sale of such products in megacities and towns has been slowly growing for the past two to three years.

Presently, many of these microbial enzymes, such as microbial rennets and other enzymes, are being imported. Hence, there is a scope for the production of enzymes such as microbial rennet, lactase, proteinases, and lipases indigenously. In the near future, the requirement for these enzymes is bound to increase by leaps and bounds, basically due to requirement of value-added dairy products in the country.

Common Tests of Milk and Cream

Milk and dairy products constitute an important item of our food. These products are very suitable for microbial growth. It thus becomes necessary to know the chemistry of milk, its spoilage, method of preservation, and different dairy products where microbes play a positive rather than negative role.

Milk is considered as a complete food and it contains proteins, fat, carbohydrates, minerals, vitamins and water. It is also a good medium for the growth of microorganisms. It is therefore, important to know the types of microorganisms present in milk, their control and use for beneficial purposes.

Milk contains relatively few bacteria when it is secreted from the udder of an healthy animal. However, during milking operations it gets contaminated from the exterior of the upper and the adjacent areas, dairy untensils, milking machines, the himds of the milkers from the soil and dust. In this way bacteria, yeasts and molds got into the milk and constitute the normal flora of milk. Thc number of contaminants added from various sources depends on the care taken to avoid contamination.

The presence of these nonpathogenic organisms in milk is not serious but if these organisms multiply quickly, they can cause spoilage of milk, such as souring or putrefaction and develop undesirable odours. Control of their multiplication in milk is therefore, very essential.

Milk may also contain pathogenic organisms, derived directly from the animal or from the surroundings. Microorganisms that are harmful and found in milk are Streptococcus cremoris, Pseudomonas sp., Mycobacterium spp. Serratia marcescens, enteric bacteria etc. Normally, milk is pasteurized before use. However, pasteurization does not kill all the bacteria; the survivors (thermodurics), depending on their initial number then multiply and if the initial number is high they cause rapid spoilage. It is imporant, therefore, that the milk be refrigerated at. around O°C soon after pasteurization to prevent the growth of these undersirable microorganisms.

Pasteurization, either at 145°F for 30 minutes or 161°P for 15'"30 seconds eliminates most of the pathogenic bacteria particularly Mycobacterium tuberculosis.

Boiling of milk destroys all microorganisms except spore formers. Sometimes, on cooling or under improper refrigeration, spores germinate and cause spoilage of boiled milk.

Detailed Descriptions: Milk Component Analysis

Analyst Performing Milk Component Test with the MilkoScan™ FT 120

This electronic instrument provides us the ability to quickly analyse milk samples for components (i.e. fat, protein, lactose, total solids and solids non-fat). MilkoScan ™ FT 120 employs the Fourier transform Infra-Red Spectrophotometer (FTIR) measuring principle. In basic terms, the instrument provides milk component results through infrared light measurement. The Foss MilkoScan ™ FT 120 can provide test results in about 30 seconds and can be calibrated for raw bovine and caprine milk, homogenized milks, cream, whey and other dairy products.

Milk Component Chemistry

Babcock Method for Fat : This is relatively inexpensive test for fat that can be performed on raw milk and a limited number of dairy products. This procedure uses sulfuric acid to "burn everything in the milk sample" except for the fat. The remaining fat in the sample is then centrifuged and determined graduated markings on specialized glassware. The procedure is limited in that individual samples can only be read to the 0.05% fat level.

Gerber Method for Fat

This is a test is close relative to the Babcock method. It is often referred to as Babcock's European cousin as this test is most common in the European dairy industry. Milk fat is separated from other milk components by adding sulfuric acid. The separation is facilitated by using amyl alcohol and centrifugation. The fat content is read directly via a special calibrated butyrometer. The procedure is limited in that individual samples can only be read to the 0.05% fat level.

Mojonnier (Ether Extract) Method

This is the chemical method of choice for total milk fat because of its inherent analytical capacity. It can provide a very precise measurement of milk fat. Because of its precision, the ether extraction procedure is the preferred chemical reference method for milk fat. Ether and alcohol are used to extract the fat from milk. The ether extract is decanted into a dry weighing dish and the ether is evaporated. The extracted fat is then dried to a constant weight and is expressed as % fat by weight.

Kjeldahl Method for Protein

This chemical method determines the nitrogen content of milk. Acid digestion of the milk sample occurs at high temperature. Nitrogen content is then converted by known formula to provide the milk protein%.

Direct Forced Air Oven Method for Total Solids

Total solids are determined by weighing milk, drying milk and weighing the dried milk residue. The sample is dried in a forced air oven for 4 hours at 100 + 1°C. The total solids content of milk is the weight of dried milk expressed as a percentage of the original milk weight.

Electronic Somatic Cell Count (ESCC) with the Fossomatic ™ Minor

Electronic instrument used in the detection of Somatic cells in milk. The milk sample is mixed with a staining solution that dyes the DNA in somatic cells. The stained sample is then pipetted into the instrument. Light is used to fluoresce the dyed somatic cells. The fluoresced cells are then counted by using a light sensitive chip.

Direct Microscopic Somatic Cell Count

The SCC is determined by viewing stained somatic cells under a microscope under oil immersion. The analysts' estimated counts are determined by using a microscopic factor based on the field of view of the microscope.

Freezing Point Depression (FPD) of Milk with the Cryoscope

This test is used to detect the adulteration of water in milk. Due to the soluble components of milk (i.e. lactose and salts), normal milk freezes slightly below the "freezing point" of pure water. The cryoscope instrument is used to determine the freezing point of milk. Small amounts of added water dilute the soluble components of milk causing the freezing point to move upward toward that of pure water.

Interpretations of cryoscope readings can be awkward. Years ago, cryoscope readings were reported with a unique measurement scale known at the Hortvet scale (°H). However, today most cryoscope results are reported in Centigrade (°C). In some cases, laboratories have inadvertently calibrated their cryoscopes to °H and reported results as °C. When interpreting cryoscope results, first determine which measurement scale (°H or °C), is being used to calibrate the lab's cryoscope. Then apply the reading to the appropriate scale. Normal comingled loads of milk should have cryoscope values between-0.517 and-0.527 °C (-0.540 and-0.550 °H).

However, normal single herd result may vary depending on a variety of conditions including (but not limited to) herd health, season of the year, mineral content of the herd's diet, animal breed and herd stress. Because of these factors, the concerns implicating adulteration of milk with water usually are noted at-0.503 °C (-0.525 °H). If a herd's cryoscope results consistently show concern, an evaluation of the herd's milk should be conducted to determine a "baseline" for that specific herd's normal cryoscope reading. The cryoscope at Regulatory Services' milk lab is calibrated in °C and all test results are reported in °C.

Standard Plate Count Bacteria (SPC) with Petrifilm®

The Petrifilm® plate is an all-in-one plating system method that provides for the Standard Plate Count (SPC) of aerobic bacteria in milk. Pre-made films of agar are inoculated with 1:100 and 1:1000 dilutions of milk. The inoculated Petrifilm® is incubated for 48 hours and the bacteria colonies are counted by the analyst.

Antibiotic Test with Charm SL®

Detection of antibiotics or other growth inhibitors in raw milk is made possible by validated growth inhibitor assays or by rapid receptor based screening processes. Acceptable test methods have been validated and approved by FDA are identified in PMO related documents. The Charm SL® is approved for the determination of antibiotics in raw cow and goat milk. The Charm SL® is a receptor assay utilizing rapid one step assay lateral flow technology. The procedure detects Beta-

Lactam drugs (Amoxicillin, Ampicillin, Ceftiofor, Cephapicin, Penicillin G) at or below safe level/tolerance.

Sediment

The sediment test determines the level of particulates present in milk and is an indicator of milk production practices and cleanliness. Sediment is determined by filtering a portion of milk by suction through a fine fibre filter. The filter is then visually examined to determine the milligrams of sediment present per gallon. In years past, the sediment test was quite common throughout the dairy industry. Improvements in farm production practices and in milk quality have diminished the need for routine sediment testing in most parts of the US.

Other Tests

The milk laboratory at Regulatory Services has the capacity to perform the milk analyses identified above. Periodically we may add new procedures. Please feel free to contact us for milk testing inquiries not identified in this document.

Milk Testing and Quality Control

Milk testing and quality control is an essential component of any milk processing industry whether small, medium or large scale. Milk being made up of 87% water is prone to adulteration by unscrupulous middlemen and unfaithful farm workers. Moreover, its high nutritive value makes it an ideal medium for the rapid multiplication of bacteria, particularly under unhygienic production and storage at ambient temperatures. We know that, in order for any processor to make good dairy products, good quality raw materials are essential. A milk processor or handler will only be assured of the quality of raw milk if certain basic quality tests are carried out at various stages of transportation of milk from the producer to the processor and finally to the consumer. There are a number of standard manuals and text books on milk quality control. However these may not be easily available to the emerging small scale to medium scale processors in Kenya.

For these reasons, the Training Programme for the Small Scale Dairy Sector under project GOK/FAO/TCP/KEN/6611, has assembled this guide on Milk Testing and Quality Control so that it may be used for training and by the private small scale dairy processors. The methods selected are simple and basic and will suffice the requirements of most milk quality control laboratories of small scale processing units. For the larger plants with bigger laboratories more tests are to be found in the bibliography at the end of this booklet.

Milk Testing And Quality Control

What is Milk Quality Control?

Milk quality control is the use of approved tests to ensure the application of approved practices, standards and regulations concerning the milk and milk products. The tests are designed to ensure that milk products meet accepted standards for chemical composition and purity as well as levels of different micro-organisms.

Why have Milk Quality Control?

Testing milk and milk products for quality and monitoring that MILK PRODUCTS, PROCESSORS and MARKETING AGENCIES adhere to accepted codes of practices costs money. There must be good reasons why we have to have a quality control system for the dairy industry in Kenya.

The reasons are:

- *To the Milk Producer:* The milk producer expects a fair price in accordance with the quality of milk she/he produces.
- *The Milk Processor:* The milk processor who pays the producer must assure himself/herself that the milk received for processing is of normal composition and is suitable for processing into various dairy products.
- *The Consumer:* The consumer expects to pay a fair price for milk and milk products of acceptable to excellent quality.
- The Public and Government Agencies. These have to ensure that the health and nutritional status of the people is protected from consumption of contaminated and sub-standard foodstuffs and that prices paid are fair to the milk producers, the milk processor and the final consumer.

All the above-is only possible through institution of a workable quality testing and assurance system conforms to national or internationally acceptable standards.

Quality Control in the Milk Marketing Chain in Kenya

i) At the farm. Quality control and assurance must begin at the farm. This is achieved through farmers using approved practices of milk production and handling; and observation of laid down regulations regarding, use of veterinary drugs on lactating animals, regulations against adulterations of milk etc.

ii) At Milk collection Centres. All milk from different farmers or bulked milk from various collecting centres must be checked for wholesomeness, bacteriological, and chemical quality.

iii) At the Dairy Factories. Milk from individual farmers or bulked milk from various collecting centres

iv) Within the Dairy Factory. Once the dairy factor has accepted the farmer milk it has the responsibility of ensuring that the milk is handled hygienically during processing. It must carry out quality assurance test to ensure that the products produced conform to specified standards as to the adequacy of effect of processes applied and the keeping quality of manufactured products. A good example is the phosphatase test used on pasteurised milk and the acidity development test done on U.H.T milk.

v) During marketing of processed products. Public Health authorities are employed by law to check the quality of food stuffs sold for public consumption and may impound substandard or contaminated foodstuffs including possible prosecution of culprits. This is done in order to protect the interest of the milk consuming public.

Techniques Used in Milk Testing and Quality Control

Milk Sampling

Accurate sampling is the first pre-requisite for fair and just quality control system. Liquid milk in cans and bulk tanks should be thoroughly mixed to disperse the milk fat before a milk sample is taken for any chemical control tests. Representative samples of packed products must be taken for any investigation on quality. Plungers and dippers me used in sampling milk from milk cans.

Sampling Milk for Bacteriological Testing

Sampling milk for bacteriological tests require a lot of care. Dippers used must have been sterilised in an autoclave or pressure cooker for at least 15mm at 120° C before hand in order not to contaminate the sample. On the spot sterilisation may be employed using 70% Alcohol swab and flaming or scaling in hot steam or boiling water for 1 minute.

Preservation of Sample

Milk samples for chemical tests: Milk samples for butterfat testing may be preserved with chemicals like Potassium dichromate (1 Tablet or ½ ml 14% solution in a ¼ litre sample bottle is adequate.) Milk samples that have been kept cooling a refrigerator or ice-box must first be warmed in water bath at 40 °C, cooled to 20°C, mixed and a sample then taken for butterfat determination. Other preservative chemicals include Sodium azid at the rate of 0.08% and Bronopol (2-

bromo-2-nitro-1,3-propanediol) used at the rate of 0.02%. If the laboratory cannot start work on a sample immediately after sampling, the sample must be cooled to near freezing point quickly and be kept cool till the work can start. If samples are to be taken in the field e.g. at a milk cooling centre, ice boxes with ice pecks are useful.

Labelling and Records Keeping

Samples must be clearly labelled with name of farmer or code number and records of dates, and places included in standard data sheets. Good records must be kept neat and in a dry place. It is desirable that milk producers should see their milk being tested, and the records should be made available to them if they so require.

Common Testing of Milk

Organoleptic Tests

The organoleptic test permits rapid segregation of poor quality milk at the milk receiving platform. No equipment is required, but the milk grader must have good sense of sight, smell and taste. The result of the test is obtained instantly, and the cost of the test are low. Milk which cannot be adequately judged organoleptically must be subjected to other more sensitive and objective tests.

Procedure:

- Open a can of milk.
- Immediately smell the milk.
- Observe the appearance of the milk.
- If still unable to make a clear judgement, taste the milk, but do not swallow it. Spit the milk sample into a bucket provided for that purpose or into a drain basin, flush with water.
- Look at the can lid and the milk can to check cleanliness.

Abnormal smell and taste may be caused by:

- Atmospheric taint (e.g. barny/cowy odour).
- Physiological taints (hormonal imbalance, cows in late lactation-spontaneous rancidity).
- Bacterial taints.
- Chemical taints or discolouring.
- Advanced acidification (pH < 6.4).

Clot on Boiling (C.O.B) Test

The test is quick and simple. It is one of the old tests for too acid milk (pH<5.8) or abnormal milk (e.g. colostral or mastitis milk). If a milk sample fails in the test, the milk must contain many acid or rennet producing microrganisms or the milk has an abnormal high percentage of proteins like colostral milk. Such milk cannot stand the heat treatment in milk processing and must therefore be rejected.

Procedure: Boil a small amount of milk in a spoon, test tube or other suitable container. If there is clotting, coagulation or precipitation, the milk has failed the test. Heavy contamination in freshly drawn milk cannot be detected, when the acidity is below 0.20-0.26% Lactic acid.

The Alcohol Test

The test is quick and simple. It is based on instability of the proteins when the levels of acid and/or rennet are increased and acted upon by the alcohol. Also increased levels of albumen (colostrum milk) and salt concentrates (mastitis) results in a positive test.

Procedure: The test is done by mixing equal amounts of milk and 68% of ethanol solution in a small bottle or test tube. (68 % Ethanol solution is prepared from 68 mls 96% (absolute) alcohol and 28 mls distilled water). If the tested milk is of good quality, there will be no coagulation, clotting or precipitation, but it is necessary to look for small lumps. The first clotting due to acid development can first be seen at 0.21-0.23% Lactic acid. For routine testing 2 mls milk is mixed with 2 mls 68% alcohol.

The Alcohol-Alizarin Test

The procedure for carrying out the test is the same as for alcohol test but this test is more informative. Alizarin is a colour indicator changing colour according to the acidity. The Alcohol Alizarin solution can be bought ready made or be prepared by adding 0.4 grammes alizarin powder to 1 litre of 61% alcohol solution.

Acidity Test

Bacteria that normally develop in raw milk produce more or less of lactic acid. In the acidity test the acid is neutralised with 0.1 N Sodium hydroxide and the amount of alkaline is measured. From this, the percentage of lactic acid can be calculated. Fresh milk contains in this test also "natural acidity" which is due to the natural ability to resist pH changes. The natural acidity of milk is 0.16-0.18%.

Apparatus:

- A porcelain dish or small conical flask
- 10 ml pipette, graduated
- 1 ml pipette

- A Burette, 0.1 ml graduations
- A glass rod for stirring the milk in the dish
- A Phenophtalein indicator solution, 0.5% in 50% Alcohol
- N Sodium hydroxide solution.

Procedure: 9 ml of the milk measured into the porcelain dish/conical flask,1 ml Phenopthalein is added and then slowly from the burret, 0.1 N Sodium hydroxide under continuous mixing, until a faint pink colour appears. The number of mls of Sodium hydroxide solution divided by 10 expresses the percentage of lactic acid.

Resazurin Test

Resazurin test is the most widely used test for hygiene and the potential keeping quality of raw milk. Resazurin is a dye indicator. Under specified conditions Resazurin is dissolved in distilled boiled water. The Resazurin solution can later be used to test the microbial activity in a given milk sample.

Resazurin can be carried out as:

- 10 min test.
- 1 hr test.
- 3 hr test.

The 10 min Resazurin test is useful and rapid, screening test used at the milk platform. The 1 hr test and 3 hr tests provide more accurate information about the milk quality, but after a fairy long time. They are usually carried out in the laboratory.

Apparatus and reagents:

- Resazurin tablets
- Test tubes with 10 mls mark
- 1 ml pipette or dispenser for Resazurin solution.
- Water bath thermostatically controlled
- Lovibond comparator with Resazurin disc 4/9.

Procedure: The solution of Resazurin as prepared by adding one tablet to 50 mls of distilled sterile water. Rasazurin solution must not be exposed to sunlight, and it should not be used for more than eight hours because it losses strength. Mix the milk and with a sanitized dipper put 10 mls milk into a sterile test tube.

Add one ml of Resazurin solution, stopper with a sterile stopper, mix gently the dye into the milk and mark the tube before the incubation in a water bath, place the test tube in a Lovibond comparator with

Resazurin disk and compare it colourimetrically with a test tube containing 10 ml milk of the same sample, but without the dye (Blank).

The Gerber Butterfat Test

The fat content of milk and cream is the most important single factor in determining the price to be paid for milk supplied by farmers in many countries.

Also, in order to calculate the correct amount of feed ration for high yielding dairy cows, it is important to know the butterfat percentage as well as well as the yield of the milk produced. Further more the butterfat percentage in the milk of individual animals must be known in many breeding programmes. Butterfat tests are also done on milk and milk products in order to make accurate adjustments of the butterfat percentage in standardised milk and milk products.

Apparatus for DF test:

- Gerber butyrameters, 0-6% or 0-8% BF
- Rubber stoppers for butyrometers
- 10.94 or 11 ml pipettes for milk
- 10 mls pippetes or dispensers for Gerber Acid
- 1 mls pippetes or dispensers for Amyl alcohol
- stands for butyrometers.

Gerber water bath Reagents:

- Gerber sulphuric acid,(1.82 g/cc)
- Amyl alcohol.

Treatment of Samples

Fresh milk at approximately 20°C should be mixed well. Samples kept cool for some days should be warmed to 40°C, mixed gently and cooled to 20°C before the testing.

Procedure: Add 10 mIs sulphuric acid to the butyrometer followed by 10.94 or 11 mls of well mixed milk. Avoid wetting of the neck of the butyrometer.

Next add 1 ml of Amyl alcohol, insert stopper and shake the butyrometer carefully until the curd dissolves and no white particles can be seen. Place the butyrometer in the water bath at 65°C and keep it there until a set is ready for centrifuging. The butyrometer must be placed in the centrifuge with the stem (scale) pointing towards the centre of the centrifuge.

Spin for 5 min. at 1100 rpm.

Remove the butyrometers from the centrifuge.

Put the butyrometers in a water bath maintained at 65°C for 3 min. before taking the reading.

(Note: When transferring the butyrometers from the centrifuge into the water bath make sure that the butyrometers are all the time held with the NECK POINTING UP).

The fat column should be read from the lowest point of the meniscus of the interface of the acid-fat to the 0-mark of the scale and read the butterfat percentage.

The butyrometers should be emptied into a special container for the very corrosive liquid of acid-milk, and the butyrometers should be washed in warm water and dried before the next use.

Appearance of the Test:

- The colour of the fat column should be straw yellow.
- The ends of the fat column should be clearly and sharply defined.
- The fat column should be free from specks and sediment.
- The water just below the fat column should be perfectly clear.
- The fat should be within the graduation.

Problems in Test Results

Curdy tests:

- Too lightly coloured or curdy fat column can be due to:
- Temperature at milk or acid or both too low.
- Acid too weak.
- Insufficient acid.
- Milk and acid not mixed thoroughly.

Charred tests:

- Darkened fat column containing black speck at the base is due to:
- Temperature of milk-acid mixture too high.
- Acid too strong.
- Milk and acid mixed too slowly.
- Too much acid used.
- Acid dropped through the milk.

The Lactometer test

Addition of water to milk can be a big problem where we have unfaithful farm workers, milk transporters and greedy milk hawkers. A few farmers may also fall victim of this illegal practice. Any buyer of milk should therefore assure himself/herself that the milk he/she purchases is wholesome and has not been adulterated. Milk has a specific gravity. When its adultered with water or other materials are added or both misdeeds are committed, the density of milk change from its normal value to abnormal.

The lactometer test is designed to detect the change in density of such adulterated milk. Carried out together with the Gerber butterfat test, it enables the milk processor to calculate the milk total solids (% TS) and solids not fat (SNF). In normal milk SNF should not be below 8.5% according to Kenya Standards(KBS No 05-l0:-1976).

Procedure: Mix the milk sample gently and pour it gently into a measuring cylinder (300-500). Let the Lactometer sink slowly into the milk. Read and record the last Lactometer degree (°L) just above the surface of the milk. If the temperature of the milk is different from the calibration temperature (Calibration temperature may be=20 0C) of the lactometer, calculate the temperature correction. For each °C above the calibration temperature add 0.2°L; for each °C below calibration temperature subtract 0.2 °L from the recorded lactometer reading.

For the calculations, use lactometer degrees, and for the conversion to density write 1.0 in front of the true lactometer reading,i.e. 1.030 g/ml. Clever people may try to adulterate milk in such a way that the lactometer cannot show the adulteration. But look to see if there is an unusual sediment from the milk at the bottom of the milk can and taste to find out if the milk is too sweet or salty to be normal. Samples of milk from individual cows often have lactometer reading outside the range of average milk, while samples of milk from herds should have readings hear the average milk, but wrong feeding, may result in low readings. Kenyan standards expects milk to have specific gravity of 1.026-1.032 g/ml which implies a Lactometer reading range of 26.0-32.0 °L. If the reading is consistently lower than expected and the milk supplier disputes any wrong doing arrange to take a genuine sample from the supplier (i.e. inspect milk right from source).

Freezing Point Determination

The freezing point of milk is regarded to be the most constant of all measurable properties of milk. A small adulteration of milk with water will cause a detectable elevation of the freezing point of milk from its normal values of-0.54°C. Since the test is accurate and sensitive to added water in milk, it is used to detect whether milk is of normal composition and adulterated.

Inhibitor Test

Milk collected from producers may contain drugs and/or pesticides residues. These when present in significant amounts in milk may inhibit the growth of lactic acid bacteria used in the manufacture of fermented milk such as Mala, cheese and Yoghurt, besides being a health hazard.

Principle of the Method: The suspected milk sample is subjected to a fermentation test with starter culture and the acidity checked after three (3) hours. The values of the titratable acidity obtained is compared with titratable acidity of a similarly treated sample which is free from any inhibitory substances.

Materials:

- test tubes
- Starter culture
- lml pipette
- water bath
- material for determination of titratable acidity.

Procedure: Three test tubes are filled with 10 ml of sample to be tested and three test tubes filled with normal milk.

All tubes are heated to 90 0C by putting them in boiling water for 3-5 minutes.

After cooling to optimum temperature of the starter culture (30,37, or 42°C), 1 ml of starter culture is added to each test tube, mixed and incubated for 3 hours.

After each hour, one test tube is from the test sample and the control sample is determined.

Assessment of results:

- If acid production in suspected sample is the same as the normal sample, then the suspect sample does not contain any inhibitory substances;
- If acid production as suspect sample is less than in the normal milk sample, then, the suspect sample contains antibiotics or other inhibitory substances.

Quality Control of Pasteurised Milk

When milk is pasteurised at 63°C for 30 min in batch pasteuriser or 72°C for 15 seconds in heat exchanger, continuous flow pasteurisers,

All Pathogenic Bacteria Are Destroyed, there by rendering milk safe for human consumption. Simultaneously various enzymes present in milk, and which might affect its flavour, are destroyed.

In order to determine whether or not milk has been adequately pasteurised, one of the enzymes normally present in milk phosphatase, is measured. A negative phosphatase result indicates that the enzyme and any pathogenic bacteria have been destroyed during pasteursation. If it is positive, it means the pasteurisation process was inadequate and the milk may not be safe for human consumption and will have a short shelf life.

- Test tubes
- 5 mls pipettes
- 1 ml pipettes
- 100 ml volumetric flask
- 500 ml volumetric flask
- water bath at 37°C.

Note: All glassware must be rinsed, cleaned, rinsed in chromic acid solution and boiled in water for 30 min.

Reagent

Buffer Solution: Is mixed by 0.75g anhydrous sodium carbonate and 1.75g Sodium bicarbonate in 500 ml distilled water.

Buffer-substrate Solution: Place 0.15 g of di-sodium paranitrophenylphosphate (the substrate) into a clean 100ml measuring cylinder.

Add the buffer solution to make to 100 ml mark.

Store this buffer-substrate solution in a refrigerator and protected against light. It should not be used after one week. Prepare a fresh stock.

Procedure: Pipette 5mls buffer-substrate solution into a test tube, stopper and warm the solution in the water bath at 37°C. Add to the test tube 1ml of the milk to be tested, stopper and mix well and place in water bath at 37°C. Prepare a blank sample from boiled milk of the same type as that undergoing the test. Incubate both the test samples and the blank sample at 37°C for 2hrs. After incubation, remove the tubes and mix them thoroughly.

Place one sample against the blank in a Lovibond comparator" All Purposes" using A.P.T.W. disc and rotate the disc until the colour of the test sample is matched and read the disc number.

Interpretation:

Disc Reading after 2 hrs incubation at 37°C	*Remarks*
0-10	Properly pasteurised
10-18	Slightly under pasteurised
18-42	UNDER PASTEURISED
> 42	NOT PASTEURISED

Common Adulterants/Contaminants in Food and Simple Screening Tests for their Detection

Adulteration in food is normally present in its most crude form, prohibited substances are either added or partly or wholly substituted. In India normally the contamination/adulteration in food is done either for financial gain or due to carelessness and lack in proper hygienic condition of processing, storing, transportation and marketing. This ultimately results that the consumer is either cheated or often become victim of diseases.

Such types of adulteration are quite common in developing countries or backward countries. However, adequate precautions taken by the consumer at the time of purchase of such produce can make him alert to avoid procurement of such food. It is equally important for the consumer to know the common adulterants and their effect on health.

Chapter 2

Chemical Constituents of Milk

Composition of Milk

Milk from different animals varies somewhat in the proportion of the different constituents. All milk contains a high proportion of water, cow's milk averaging about 87 per cent. The milk from different breeds of cattle varies considerably in composition. The milk of individual cows of the same breed also varies in the proportion of the different constituents. This may be partly due to environment, partly to inheritance, and partly to individuality. Milk may vary particularly in fat content from one milking to the next. The percentage of fat increases during the milking period; that is, the first milk obtained is not so rich in fat as the last portions of milk.

The fat varies for different breeds from about 3.5 per cent for Holstein to about 5 per cent for Guernsey and Jersey.

The protein varies from about 3.3 per cent for Holstein to 4.0 per cent for Guernsey and Jersey. It parallels the fat content somewhat, being highest in the breeds having a high fat content and lowest in those having a low fat content.

The lactose does not vary so much with the different breeds as the fat and protein. The lactose content is from about 4.65 to 5 per cent.

The ash varies from about 0.68 to 0.75 for the different breeds.

The constituents of milk that are most important in food preparation are enzymes, vitamins, pigments, salts, sugar, fat, and proteins. Enzymes. The enzymes of cow's milk are reported as follows by Rogers; proteinases, lactase, diastase, lipase, salolase, catalase, peroxidase, and aldehydrase. Rogers states that the proteolytic enzyme, galactase, brings about slow decomposition of milk proteins into peptones, amino acids, and ammonia. Vitamins. All the vitamins recognized at the present time are contained in milk. Some are present in comparatively large and others in smaller amounts.

Pigments. The appearance of milk is white. This is due to light rays reflected by the colloidally dispersed constituents of the milk, the calcium caseinate, and calcium phosphate.

Milk contains two classes of yellow or orange pigments. The water-soluble pigment, which imparts a yellow colour with a green fluorescence to the whey of milk, was formerly called lactochrome. A name recently suggested for this pigment is lactoflavin. It is regarded as one flavin of a specific group, collectively to be called lyochromes. It is possible that lacto-flavin is composed of more than one pigment. Rogers says "lactoflavin forms compounds with saccharides, proteins, and purines (uric acid). These compounds possibly either occur naturally in milk or readily form during isolations, thus accounting for the several lactoflavins isolated from milk." It is probable that the pigment lactoflavin is one of the five fractions of vitamin G (B2). Milk is relatively rich in this vitamin.

A fat-soluble pigment, carotene, found in the fat gives the milk a more or less yellow tinge, which is more pronounced as the fat particles become more concentrated and form cream The group of pigments called caroti-noids, which include carotene, xanthophyll, and related pigments, has been described in the chapter on fruits and vegetables. The chief pigment of butter fat is the carotene, but little xanthophyll being found. The depth of colour depends upon the amount of pigment present.

The colour of carotene in solution varies from yellow to orange and to a deep red-orange as the concentration increases. The amount of carotene found in the butter fat depends upon the extent of carotene in the food of the cow. Green grasses, hay cured to retain its green colour, green corn, and carrots are rich in carotene. The carotene content of milk fat is less rich during the winter months, if the food of the cow is poor in carotene during this period. Only in cow's milk is the fat extensively pigmented. With the exception of the fat of human milk, which is often pigmented, the fat of the milk of other animals is either devoid of or contains little pigment.

Salts. Milk contains salts of potassium, sodium, magnesium, calcium, phosphates, chlorides, and citrates. Traces of sulfates and carbonates are found. Iron is present in small amount. Iodides are also found in small amounts, the amount being greater in some localities than in others. Iodides may be easily transmitted from the feed to the milk. Supplee and Bellis have found copper to average about 0.52 part per million in freshly drawn milk.

Brickner has reported that milk contains 3.6 to 5.6 parts of zinc per million parts of milk. Manganese in normal milk averages 0.02 to 0.06 parts per million. The greater part of the sulfur is found in the milk proteins. Barger and Coyne state that part of the sulfur in milk is found in the amino acids methionine and cystine of the proteins, but that all of the sulfur is not accounted for.

The salts of milk are found in milk in solution, in the colloidal state, and in combination with the proteins. The exact chemical combinations of the different salts in the milk are not fully determined. For this reason different authorities report different salt combinations. Thus formulas imply definite combination, whereas there is a complex salt equilibrium in milk, which has not been satisfactorily worked out. The citrates, the combinations of which may be trisodium and tripotassium citrate, tricalcium and trimagnesium citrate, are probably entirely in solution. The possible chlorides of potassium, sodium, and calcium are also in solution. Some authorities believe that the phosphates are present chiefly as dicalcium phosphate, CaHPO4 others believe that tricalcium phosphate, Ca3(PO4)2, is the principal phosphate combination. The phosphates are partly in solution, with the greater portion in colloidal dispersion. When the particles of dicalcium phosphate are heated they become aggregated and partially precipitated. Calcium and magnesium are in combination with the casein to form calcium and magnesium caseinates. Zoller states that there may be traces of sodium and potassium caseinates.

Lactose. The solubility of lactose and its properties may be found in the chapter on sugar. It is caramelized by heat at rather low temperatures.

Fat. Butter fat is composed of glycerol and fatty acids. Fatty acids of both the saturated and unsaturated series are present. The relative percentage of the unsaturated fatty acids varies with the feed, averaging higher in summer than in winter. Dean and Hilditch state that the oleic and linoleic acids increase by 4 per cent (mols), with a parallel diminution in butyric and stearic acids when cows return to pasture. They also report that with increased age of the cow the unsaturated acids increase at the expense of palmitic acid, which was lowered from 29 to 22-23 per cent by weight of the total fat in the four years of observations made on milk from the same group of cows. The saturated fatty acids are butyric, caproic, caprylic, capric, lauric, myristic, plamitic, and stearic.

Arachadonic acid has been reported absent by some investigators. Butter fat contains a higher proportion of the first-named saturated acids than other food fats. The first ones in the series are quite volatile with steam, the volatility decreasing with increase in molecular weight of the acid. Hence, when butter is heated for several minutes, the percentage of the lower saturated fatty acids may be decreased. The unsaturated fatty acids include oleic, linoleic, and arachidic.

Formerly it was thought that the particular flavour of butter was due to the greater proportion of the lower saturated fatty acids, but now it is known that butters with satisfactory flavour and aroma (Michaelian, Farmer, and Hammer) contain considerable quantities of acetylmethyl-carbinol plus diacetyl. The acetylmethylcarbinol in a pure condition is odorless, but by bacterial action it is changed to diacetyl, which in concentrations of 0.0002 to 0.0004 per cent in or added to neutral butter gives a characteristic aroma.

Size of Fat Globules

The fat in milk is found in small globules. These fat globules which are microscopic in size are suspended in the milk. They vary in size from 0.10μ to 22.0μ. Most of the globules are less than 10μ and average about 3μ in diameter. The size of the fat globules varies (1) with different breeds, being larger in milk from Jersey and Guernsey than in milk from other breeds, (2) with the lactation period, decreasing in size with length of the lactation period, and (3) with feed. Dry feed tends to decrease the size of the globules, succulent feed to increase their size.

Creaming

The specific gravity of the fat globules is less than that of the fluid of the milk. Hence there is a tendency for them to rise to the surface to form a cream layer. The extent of creaming depends upon several factors, such as the size of the fat globules, temperature of the milk, acidity, physical state of the fat, etc. Creaming occurs more rapidly in milk when the fat globules are quite large than when they are smaller. In rising to the top of the milk the globules of fat clump together, and this clumping increases the tendency for them to rise to the surface of the milk. As the larger clumps rise they may carry many of the smaller globules to the surface. Therefore, clumping not only aids the completeness of creaming but also the rate, for the rate is more rapid when the fat particles clump quickly. Rogers states that the factors affecting clumping are "temperature, the acidity, the fat content and its degree of dispersion, the degree of agitation, and the fluidity of the system.

The fat content, the degree of agitation and the fluidity of the system determine the probability of collisions of the globules." The tendency of the fat globules to clump is greatest when the milk is cooled rapidly to 7° to 8°C, but if the fat globules become solid at the low temperature before creaming is allowed to take place the rate of creaming is retarded. The temperature that favours clumping is also best for whipping the cream. Most cream is separated from the milk by mechanical means. The fat particles left in the milk after separating the cream are those less than 1u and those between 1 and 2u in diameter.

Butter

The fat globules in cow's milk are suspended in the milk and thus do not form a permanent emulsion, though they may be so reduced in size by homogenization that they form a permanent emulsion. Of the different theories formulated for explaining the manner in which emulsions are stabilized the adsorption film theory is usually connected with milk. Substances that lower surface tension tend to collect at the interface between two non-miscible systems. Proteins tend to lower the surface tension, hence tend to collect at the interface.

The layer or film around the fat globules probably consists of adsorbed calcium caseinate, with some lactal-bumin, globulin, and calcium phosphate. This membrane may be weakened or broken in various ways. When milk is heated slowly, the membrane surrounding some of the fat globules may be broken and a number of globules may coalesce. Sometimes milk that has been heated and then cooled has a more oily appearance due to this coalescing of the fat globules. The membrane surrounding the fat particles may be broken by mechanical agitation. Formation of butter in churning is brought about in this manner.

There are Two Theories Regarding Butter Formation: One is that the emulsion is reversed from the type found in the milk and that the butter is a water-in-fat emulsion. The other view is that butter is formed by packing the fat globules into a compact mass, and that water and air are enmeshed during this process. Temperature and the formation of a foam are both important in churning. At a temperature above 65°C. there is no aggregation of fat particles. Below 65° the tendency to clump increases and is at a maximum at 7° to 8°C. A favourable temperature for butter formation is below 24° and above 10°C. At temperatures below 4°C. the fat globules do not adhere to each other and aggregation does not take place. At temperatures above the melting point of butter, no butter is formed.

The fat particles tend to clump at the liquid/air interface, so that air beaten in during the churning process accelerates clumping of the fat. Butter may be churned from sweet or sour cream. The flavour of butter from the sweet cream is milder and different from that of the sour-cream butter. Many housekeepers prefer the sweet-cream butter for table use and the sour-cream butter for cooking.

The adsorbed films surrounding the fat globules may also be destroyed by the addition of acid or alkali. It is by these methods that the fat is set free for a quantitative determination. In the Babcock test, acid is used for liberating the fat globules; in the Hoyberg test alkali is used. The Bab-cock, or some modification of it, is the one usually employed in estimating the fat content of milk and cream.

Protein

The chief proteins found in milk in order of their decreasing amounts are casein, lactalbumin, and lactoglobulin.

Casein

Casein belongs to the group of phosphoproteins. The form in which the phosphorus exists in the casein is not definitely known, but it is believed to be present in the form of combined phosphoric acid. Casein forms about 3 per cent of cow's milk. At its isoelectric point, which is pH 4.6, casein is nearly insoluble in water. Casein is amphoteric and forms salts with acids and alkalies. Fresh milk has a reaction of about pH 6.6, so that the casein is present in the milk as salts of bases and is found as calcium and magnesium caseinates. All the alkali caseinates are soluble in water, though the salts of the alkaline earths are less soluble than the alkali ones. Loeb states that below pH 4.6 the casein chloride, casein acetate, and casein lactate are very soluble in water, but casein sulfate and casein oxalate are difficultly soluble.

According to Zoller, pure casein when heated in water begins to imbibe water at 80° to 90°C. and becomes plastic. In this form it can be molded and shaped. Upon cooling it becomes very hard.

Casein can be precipitated from milk by bringing the milk to the isoelectric point of casein. Coagulation of casein will be considered later.

Lactalbumin

The proportion of lactalbumin in milk is much lower than that of casein. It forms about 0.50 per cent of cow's milk. Its isoelectric point is pH 4.55. Since the reaction of fresh milk is about pH 6.6, it is on the alkaline side of the isoelectric point of lactalbumin. Thus is it possible that the lactalbumin is found combined as salts of bases, such as calcium and magnesium albuminates. Osborne and Wakeman think it is uncombined. Lactalbumin is soluble in water, and is coagulated by heating in solution to a temperature of about 70°C. Coagulation may not be complete at this temperature. Palmer states that the lactalbumin is more highly dispersed than the other colloidal constituents of the milk.

Lactoglobulin

Lactoglobulin occurs in milk in very small quantities, about 0.05 per cent of cow's milk. Lactoglobulin is insoluble in distilled water, but it is soluble in dilute solutions of strong bases or acids, and in dilute salt solutions. It is coagulable by heat.

Physical and Chemical Properties of Milk

The role of milk in nature is to nourish and provide immunological protection for the mammalian young. Milk has been a food source for humans since prehistoric times; from human, goat, buffalo, sheep, yak, to the focus of this section-domesticated cow milk (genus Bos). Milk and honey are the only articles of diet whose sole function in nature is food. It is not surprising, therefore, that the nutritional value of milk is high. Milk is also a very complex food with over 100,000 different molecular species found. There are many factors that can affect milk composition such as breed variations (see introduction, cow to cow variations, herd to herd variations-including management and feed considerations, seasonal variations, and geographic variations. With all this in mind, only an approximate composition of milk can be given:

- 87.3% water (range of 85.5%-88.7%)
- 3.9 % milkfat (range of 2.4%-5.5%)
- 8.8% solids-not-fat (range of 7.9-10.0%):
  - — protein 3.25% (3/4 casein)
  - — lactose 4.6%
  - — minerals 0.65%-Ca, P, citrate, Mg, K, Na, Zn, Cl, Fe, Cu, sulfate, bicarbonate, many others
  - — acids 0.18%-citrate, formate, acetate, lactate, oxalate
  - — enzymes-peroxidase, catalase, phosphatase, lipase
  - — gases-oxygen, nitrogen
  - — vitamins-A, C, D, thiamine, riboflavin, others.

The following terms are used to describe milk fractions:

- Plasma = milk-fat (skim milk)
- Serum = plasma-casein micelles (whey)
- solids-not-fat (SNF) = proteins, lactose, minerals, acids, enzymes, vitamins
- Total Milk Solids = fat + SNF.

Not only is the composition important in determining the properties of milk, but the physical structure must also be examined. Due to its role in nature, milk is in a liquid form. This may seem curious if one takes into consideration the fact that milk has less water than most fruits and vegetables. Milk can be described as:

- an oil-in-water emulsion with the fat globules dispersed in the continuous serum phase
- a colloid suspension of casein micelles, globular proteins and lipoprotein partilcles
- a solution of lactose, soluble proteins, minerals, vitamins other components.

Looking at milk under a microscope, at low magnification (5X) a uniform but turbid liquid is observed.

At 500X magnification, spherical droplets of fat, known as fat globules, can be seen. At even higher magnification (50,000X), the casein micelles can be observed. The main structural components of milk, fat globules and casein micelles, will be examined in more detail later.

Milk Lipids-Chemical Properties

The fat content of milk is of economic importance because milk is sold on the basis of fat. Milk fatty acids originate either from microbial activity in the rumen, and transported to the secretory cells via the blood and lymph, or from synthesis in the secretory cells.

The main milk lipids are a class called triglycerides which are comprised of a glycerol backbone binding up to three different fatty acids. The fatty acids are composed of a hydrocarbon chain and a carboxyl group. The major fatty acids found in milk are:

Long Chain:

- C14-myristic 11%
- C16-palmitic 26%
- C18-stearic 10%
- C18:1-oleic 20%.

Short chain (11%):

- C4-butyric*
- C6-caproic
- C8-caprylic
- C10 – capric.

* butyric fatty acid is specific for milk fat of ruminant animals and is resposible for the rancid flavour when it is cleaved from glycerol by lipase action.

Saturated fatty acids (no double bonds), such as myristic, palmitic, and stearic make up two thirds of milk fatty acids. Oleic acid is the most abundant unsaturated fatty acid in milk with one double bond. While the cis form of geometric isomer is the most common found in nature, approximately 5% of all unsaturated bonds are in the trans position as a result of rumen hydrogenation.

Triglycerides account for 98.3% of milkfat. The distribution of fatty acids on the triglyceride chain, while there are hundreds of different combinations, is not random.

The fatty acid pattern is important when determining the physical properties of the lipids. In general, the SN1 position binds mostly longer carbon length fatty acids, and the SN3 position binds mostly shorter carbon length and unsaturated fatty acids. For example:

- C4-97% in SN3
- C6-84% in SN3
- C18-58% in SN1.

The small amounts of mono-, diglycerides, and free fatty acids in fresh milk may be a product of early lipolysis or simply incomplete synthesis. Other classes of lipids include phospholipids (0.8%) which are mainly associated with the fat globule membrane, and cholesterol (0.3%) which is mostly located in the fat globule core.

Milk Lipids-Physical Properties

The physical properties of milkfat can be summerized as follows:

- density at 20° C is 915 kg m(-3)*
- refractive index (589 nm) is 1.462 which decreases with increasing temperature
- solubility of water in fat is 0.14% (w/w) at 20° C and increases with increasing temperature
- thermal conductivity is about 0.17 J m(-1) s(-1) K(-1) at 20° C
- specific heat at 40° C is about 2.1kJ kg(-1) K(-1)
- electrical conductivity is <10(-12) ohm(-1) cm(-1)
- dielectric constant is about 3.1.

*the brackets around numbers denote superscript

At room temperature, the lipids are solid, therefore, are correctly referred to as "fat" as opposed to "oil" which is liquid at room temperature. The melting points of individual triglycerides ranges from-75° C for tributyric glycerol to 72° C for tristearin. However, the final melting point of milkfat is at 37° C because higher melting triglycerides dissolve in the liquid fat. This temperature is significant because 37° C is the body temperature of the cow and the milk would need to be liquid at this temperature. The melting curves of milkfat are complicated by the diverse lipid composition:

- trans unsaturation increases melting points
- odd-numbered and branched chains decrease melting points.

Crystallization of milkfat largely determines the physical stability of the fat globule and the consistency of high-fat dairy products, but crystal behaviour is also complicated by the wide range of different triglycerides. There are four forms that milkfat crystals can occur in; alpha, ß, ß' 1, and ß' 2, however, the alpha form is the least stable and is rarely observed in slowly cooled fat.

Milkfat Structure-Fat Globules

More than 95% of the total milk lipid is in the form of a globule ranging in size from 0.1 to 15 um in diameter. These liquid fat droplets are covered by a thin membrane, 8 to 10 nm in thickness, whose properties are completely different from both milkfat and plasma. The native fat globule membrane (FGM) is comprised of apical plasma membrane of the secretory cell which continually envelopes the lipid droplets as they pass into the lumen. The major components of the native FGM, therefore, is protein and phospholipids. The phospholipids are involved in the oxidation of milk. There may be some rearrangement of the membrane after release into the lumen as amphiphilic substances from the plasma adsorb onto the fat globule and parts of the membrane dissolve into either the globule core or the serum. The FGM decreases the lipid-serum interface to very low values, 1 to 2.5 mN/m, preventing the globules from immediate flocculation and coalescence, as well as protecting them from enzymatic action.

It is well known that if raw milk or cream is left to stand, it will separate. Stokes' Law predicts that fat globules will cream due to the differences in densities between the fat and plasma phases of milk. However, in cold raw milk, creaming takes place faster than is predicted from this fact alone. IgM, an immunoglobulin in milk, forms a complex with lipoproteins.

This complex, known as cryoglobulin precipitates onto the fat globules and causes flocculation. This is known as cold agglutination. As fat globules cluster, the speed of rising increases and sweeps up the smaller globules with them. The cream layer forms very rapidly, within 20 to 30 min., in cold milk.

Homogenization of milk prevents this creaming by decreasing the diameter and size distribution of the fat globules, causing the speed of rise to be similar for the majority of globules. As well, homogenization causes the formation of a recombined membrane which is much similar in density to the continuous phase.

Recombined membranes are very different than native FGM. Processing steps such as homogenization, decreases the average diameter of fat globule and significantly increases the surface area. Some of the native FGM will remain adsorbed but there is no longer enough of it to cover all of the newly created surface area. Immediately after disruption of the fat globule, the surface tension raises to a high level of 15 mN/m and amphiphilic molecules in the plasma quickly adsorb to the lipid droplet to lower this value. The adsorbed layers consist mainly of serum proteins and casein micelles.

Fat Destabilization

While homogenization is the principal method for acheiving stabilization of the fat emulsion in milk, fat destabilization is necessary for structure formation inbutter, whipping cream and ice cream. Fat destabilization refers to the process of clustering and clumping (partial coalescence) of the fat globules which leads to the development of a continuous internal fat network or matrix structure in the product. Fat destabilization (sometimes "fat agglomeration") is a general term that describes the summation of several different phenomena. These include:

Coalescence: an irreversible increase in the size of fat globules and a loss of identity of the coalescing globules;

Flocculation: a reversible (with minor energy input) agglomeration/clustering of fat globules with no loss of identity of the globules in the floc; the fat globules that flocculate; they can be easily redispersed if they are held together by weak forces, or they might be harder to redisperse to they share part of their interfacial layers;

Partial coalescence: an irreversible agglomeration/clustering of fat globules, held together by a combination of fat crystals and liquid fat, and a retention of identity of individual globules as long as the crystal structure is maintained (i.e., temperature dependent, once the crystals melt, the cluster coalesces). They usually come together in a shear field, as in whipping, and it is envisioned that the crystals at the surface of the droplets are responsible for causing colliding globules to stick together, while the liquid fat partially flows between they and acts as the "cement". Partial coalescence dominates structure formation in whipped, aerated dairy emulsions, and it should be emphasized that crystals within the emulsion droplets are responsible for its occurrence.

Milk Lipids-Functional Properties

Like all fats, milkfat provides lubrication. They impart a creamy mouth feel as opposed to a dry texture. Butter flavour is unique and is derived from low levels of short chain fatty acids. If too many short chain fatty acids are hydrolyzed (separated) from the triglycerides, however, the product will taste rancid. Butter fat also acts as a reservoir for other flavours, especially in aged cheese. Fat globules produce a 'shortening' effect in cheese by keeping the protein matrix extended to give a soft texture. Fat substitutes are designed to mimic the globular property of milk fat. The spreadable range of butter fat is 16-24° C. Unfortunately butter is not spreadable at refrigeration temperatures. Milk fat provides energy (1g = 9 cal.), and nutrients (essential fatty acids, fat soluble vitamins).

Milk Proteins

Introduction and Review

The primary structure of proteins consists of a polypeptide chain of amino acids residues joined together by peptide linkages, which may also be cross-linked by disulphide bridges. Amino acids contain both a weakly basic amino group, and a weakly acid carboxyl group both connected to a hydrocarbon chain, which is unique to different amino acids.

The three-dimensional organization of proteins, or conformation, also involves secondary, tertiary, and quaternary structures. The secondary structure refers to the spatial arrangement of amino acid residues that are near one another in the linear sequence. The alpha-helix and ß-pleated sheat are examples of secondary structures arising from regular and periodic steric relationships.

The tertiary structure refers to the spatial arrangement of amino acid residues that are far apart in the linear sequence, giving rise to further coiling and folding. If the protein is tightly coiled and folded into a somewhat spherical shape, it is called a globular protein. If the protein consists of long polypeptide chains which are intermolecularly linked, they are called fibrous proteins. Quaternary structure occurs when proteins with two or more polypeptide chain subunits are associated.

Caseins

The casein content of milk represents about 80% of milk proteins. The principal casein fractions are alpha(s1) and alpha(s2)-caseins, ß-casein, and kappa-casein. The distinguishing property of all caseins is their low solubility at pH 4.6. The common compositional factor is that caseins are conjugated proteins, most with phosphate group(s) esterified to serine residues. These phosphate groups are important to the structure of the casein micelle. Calcium binding by the individual caseins is proportional to the phosphate content.

The conformation of caseins is much like that of denatured globular proteins. The high number of proline residues in caseins causes particular bending of the protein chain and inhibits the formation of close-packed, ordered secondary structures. Caseins contain no disulfide bonds. As well, the lack of tertiary structure accounts for the stability of caseins against heat denaturation because there is very little structure to unfold. Without a tertiary structure there is considerable exposure of hydrophobic residues. This results in strong association reactions of the caseins and renders them insoluble in water.

Within the group of caseins, there are several distinguishing features based on their charge distribution and sensitivity to calcium precipitation:

Alpha(s1)-casein: (molecular weight 23,000; 199 residues, 17 proline residues) Two hydrophobic regions, containing all the proline residues, separated by a polar region, which contains all but one of eight phosphate groups. It can be precipitated at very low levels of calcium.

Alpha(s2)-casein: (molecular weight 25,000; 207 residues, 10 prolines) Concentrated negative charges near N-terminus and positive charges near C-terminus. It can also be precipitated at very low levels of calcium.

B-casein: (molecular weight 24,000; 209 residues, 35 prolines) Highly charged N-terminal region and a hydrophobic C-terminal region. Very amphiphilic protein acts like a detergent molecule. Self association is temperature dependant; will form a large polymer at 20° C but not at 4° C. Less sensitive to calcium precipitation.

Kappa-casein: (molecular weight 19,000; 169 residues, 20 prolines) Very resistant to calcium precipitation, stabilizing other caseins. Rennet cleavage at the Phe105-Met106 bond eliminates the stabilizing ability, leaving a hydrophobic portion, para-kappa-casein, and a hydrophilic portion called kappa-casein glycomacropeptide (GMP), or more accurately, caseinomacropeptide (CMP).

Structure: The Casein Micelle

Most, but not all, of the casein proteins exist in a colloidal particle known as the casein micelle. Its biological function is to carry large amounts of highly insoluble CaP to mammalian young in liquid form and to form a clot in the stomach for more efficient nutrition. Besides casein protein, calcium and phosphate, the micelle also contains citrate, minor ions, lipase and plasmin enzymes, and entrapped milk serum. These micelles are rather porous structures, occupying about 4 ml/g and 6-12% of the total volume fraction of milk. The "casein sub-micelle" model has been prominent for the last several years, and is illustrated and described with the following link, but there is not universal acceptance of this model, and mounting research evidence to suggest that there is not a defined sub-micellar structure to the micelle at all. Another model of a more open structure is also defined with the following link.

In the submicelle model, it is thought that there are small aggregates of whole casein, containing 10 to 100 casein molecules, called submicelles. It is thought that there are two different kinds of submicelle; with and without kappa-casein. These submicelles contain a hydrophobic core and are covered by a hydrophilic coat which is at least partly comprised of the polar moieties of kappa-casein. The hydrophilic CMP of the kappa-casein exists as a flexible hair.

The open model also suggests there are more dense and less dense regions within the midelle, but there is less of a well-defined structure. In this model, calcium phosphate nanoclusters bind caseins and provide for the differences in density within the casein micelle.

Colloidal calcium phosphate (CCP) acts as a cement between the hundreds or even thousands of submicelles that form the casein micelle. Binding may be covalent or electrostatic. Submicelles rich in kappa-casein occupy a surface position, whereas those with less are buried in the interior. The resulting hairy layer, at least 7 nm thick, acts to prohibit further aggregation of submicelles by steric repulsion. The casein micelles are not static; there are three dynamic equilibria between the micelle and its surroundings:

- the free casein molecules and submicelles
- the free submicelles and micelles
- the dissoved colloidal calcium and phosphate.

The following factors must be considered when assessing the stability of the casein micelle:

Role of Ca++: More than 90% of the calcium content of skim milk is associated in some way or another with the casein micelle. The removal of Ca++ leads to reversible dissociation of ß-casein without micellular disintegration. The addition of Ca++ leads to aggregation.

H Bonding: Some occurs between the individual caseins in the micelle but not much because there is no secondary structure in casein proteins.

Disulphide Bonds: alpha(s1) and ß-caseins do not have any cysteine residues. If any S-S bonds occur within the micelle, they are not the driving force for stabilization.

Hydrophobic Interactions: Caseins are among the most hydrophobic proteins and there is some evidence to suggest they play a role in the stability of the micelle. It must be remembered that hydrophobic interactions are very temperature sensitive.

Electrostatic Interactions: Some of the subunit interactions may be the result of ionic bonding, but the overall micellar structure is very loose and open.

Van der Waals Forces: No sucess in relating these forces to micellular stability.

Steric stabilization: As already noted, the hairy layer interferes with interparticle approach.

There are several factors that will affect the stability of the casein micelle system:

Salt content: affects the calcium activity in the serum and calcium phosphate content of the micelles.

pH: lowering the pH leads to dissolution of calcium phosphate until, at the isoelectric point (pH 4.6), all phosphate is dissolved and the caseins precipitate.

Temperature: at 4° C, beta-casein begins to dissociate from the micelle, at 0° C, there is no micellar aggregation; freezing produces a precipitate called cryo-casein.

Heat Treatment: whey proteins become adsorbed, altering the behaviour of the micelle.

Dehydration: by ethanol, for example, leads to aggregation of the micelles.

When two or more of these factors are applied together, the effect can also be additive.

Casein Micelle Aggregation

Caseins are able to aggregate if the surface of the micelle is reactive. The Schmidt model further illustrates this.

Although the casein micelle is fairly stable, there are four major ways in which aggregation can be induced:

1. chymosin-rennet or other proteolytic enzymes as in Cheese manufacturing

2. acid
3. heat
4. age gelation.

Enzyme Coagulation

Chymosin, or rennet, is most often used for enzyme coagulation. During the primary stage, rennet cleaves the Phe(105)-Met(106) linkage of kappa-casein resulting in the formation of the soluble CMP which diffuses away from the micelle and para-kappa-casein, a distinctly hydrophobic peptide that remains on the micelle. The patch or reactive site, as illustrated in the above image, that is left on the micelles after enzymatic cleavage is necessary before aggregation of the paracasein micelles can begin.

During the secondary stage, the micelles aggregate. This is due to the loss of steric repulsion of the kappa-casein as well as the loss of electrostatic repulsion due to the decrease in pH. As the pH approaches its isoelectric point (pH 4.6), the caseins aggregate. The casein micelles also have a strong tendency to aggregate because of hydrophobic interactions. Calcium assists coagulation by creating isoelctric conditions and by acting as a bridge between micelles. The temperature at the time of coagulation is very important to both the primary and secondary stages. With an increase in temperature up to 40° C, the rate of the rennet reaction increases. During the secondary stage, increased temperatures increase the hydrophobic reaction. The tertiary stage of coagulation involves the rearrangement of micelles after a gel has formed. There is a loss of paracasein identity as the milk curd firms and syneresis begins.

Acid Coagulation

Acidification causes the casein micelles to destabilize or aggregate by decreasing their electric charge to that of the isoelectric point. At the same time, the acidity of the medium increases the solubility of minerals so that organic calcium and phosphorus contained in the micelle gradually become soluble in the aqueous phase. Casein micelles disintegrate and casein precipitates. Aggregation occurs as a result of entropically driven hydrophobic interactions.

Heat

At temperatures above the boiling point casein micelles will irreversibly aggregate. On heating, the buffer capacity of milk salts change, carbon dioxide is released, organic acids are produced, and tricalcium phophate and casein phosphate may be precipitated with the release of hydrogen ions.

Age Gelation

Age gelation is an aggregation phenomenon that affects shelf-stable, sterilized dairy products, such as concentrated milk and UHT milk products. After weeks to months storage of these products, there is a sudden sharp increase in viscosity accompanied by visible gelation and irreversible aggregation of the micelles into long chains forming a three-dimensional network. The actual cause and mechanism is not yet clear, however, some theories exist:

1. Proteolytic breakdown of the casein: bacterial or native plasmin enzymes that are resistant to heat treatment may lead to the formation of a gel
2. Chemical reactions: polymerization of casein and whey proteins due to Maillard type or other chemical reactions
3. Formation of kappa-casein-ß-lactoglobulin complexes

An excellent source of information on casein micelle stability can be found in Walstra.

Whey Proteins

The proteins appearing in the supernatant of milk after precipitation at pH 4.6 are collectively called whey proteins. These globular proteins are more water soluble than cascins and are subject to heat denaturation. Native whey proteins have good gelling and whipping properties. Denaturation increases their water holding capacity. The principle fractions are ß-lactoglobulin, alpha-lactalbumin, bovine serum albumin (BSA), and immunoglobulins (Ig).

ß-Lactoglobulins: (MW-18,000; 162 residues) This group, including eight genetic variants, comprises approximately half the total whey proteins. ß-Lactoglobulin has two internal disulfide bonds and one free thiol group. The conformation includes considerable secondary structure and exists naturally as a noncovalent linked dimer. At the isoelectric point (pH 3.5 to 5.2), the dimers are further associated to octamers but at pH below 3.4, they are dissociated to monomers.

Alpha-Lactalbumins: (MW-14,000; 123 residues) These proteins contain eight cysteine groups, all involved in internal disulfide bonds, and four tryptophan residues. alpha-Lactalbumin has a highly ordered secondary structure, and a compact, spherical tertiary structure. Thermal denaturation and pH <4.0 results in the release of bound calcium.

Enzymes

Enzymes are a group of proteins that have the ability to catalyze chemical reactions and the speed of such reactions. The action of enzymes is very specific. Milk contains both indigenous and exogenous enzymes. Exogenous enzymes mainly consist of heat-stable enzymes produced by psychrotrophic bacteria: lipases, and proteinases. There are many indigenous enzymes that have been isolated from milk. The most significant group are the hydrolases:

- lipoprotein lipase
- plasmin
- alkaline phosphatase.

Lipoprotein lipase (LPL): A lipase enzyme splits fats into glycerol and free fatty acids. This enzyme is found mainly in the plasma in association with casein micelles. The milkfat is protected from its action by the FGM. If the FGM has been damaged, or if certain cofactors (blood serum lipoproteins) are present, the LPL is able to attack the lipoproteins of the FGM. Lipolysis may be caused in this way.

Plasmin: Plasmin is a proteolytic enzyme; it splits proteins. Plasmin attacks both ß-casein and alpha(s2)-casein. It is very heat stable and responsible for the development of bitterness in pasteurized milk and UHT processed milk. It may also play a role in the ripening and flavour development of certain cheeses, such as Swiss cheese.

Alkaline phosphatase: Phosphatase enzymes are able to split specific phosporic acid esters into phosphoric acid and the related alcohols. Unlike most milk enzymes, it has a pH and temperature optima differing from physiological values; pH of 9.8. The enzyme is destroyed by minimum pasteurization temperatures, therefore, a phosphatase test can be done to ensure proper pasteurization.

Lactose

Lactose is a disaccharide (2 sugars) made up of glucose and galactose (which are both monosaccharides). Because of the anomeric carbon on the right side of the structure below, lactose can exist as two isomers, alpha, as shown, or beta, in which the hydroxyl on the anomeric carbon would point up on the ring structure shown below. It comprises 4.8 to 5.2% of milk, 52% of milk SNF, and 70% of whey solids. It is not as sweet as sucrose. When lactose is hydrolyzed by ß-D-galactosidase (lactase), an enzyme that splits these monosaccharides, the result is increased sweetness, and depressed freezing point.

One of its most important functions is its utilization as a fermentation substrate. Lactic acid bacteria produce lactic acid from lactose, which is the beginning of many fermented dairy products. Because of their ability to metabolize lactose, they have a competitive advantage over many pathogenic and spoilage organisms.

Some people suffer from lactose intolerance; they lack the lactase enzyme, hence they cannot digest lactose, or dairy products containing lactose. Crystallization of lactose occurs in an alpha form which commonly takes a tomahawk shape. This results in the defect called sandiness. Lactose is relatively insoluble which is a problem in many dairy products, ice cream, sweetened condensed milk. In addition to lactose, fresh milk contains other carbohydrates in small amounts, including glucose, galactose, and oligosaccharides.

Vitamins

Vitamins are organic substances essential for many life processes. Milk includes fat soluble vitamins A, D, E, and K. Vitamin A is derived from retinol and ß-carotene. Because milk is an important source of dietary vitamin A, fat reduced products which have lost vitamin A with the fat, are required to supplement the product with vitamin A.

Milk is also an important source of dietary water soluble vitamins:

- B1-thiamine
- B2-riboflavin
- B6-pyridoxine
- B12-cyanocobalamin
- niacin
- pantothenic acid.

There is also a small amount of vitamin C (ascorbic acid) present in raw milk but it is an insignificant amount relative to human needs and is quite heat-labile: about 20% is destroyed by pasteurization.

The vitamin content of fresh milk is given below:

Vitamin	*Contents per litre*
A (ug RE)	400
D (IU)	40
E (ug)	1000
K (ug)	50
B1 (ug)	450
B2 (ug)	1750
Niacin (ug)	900
B6 (ug)	500
Pantothenic acid (ug)	3500
Biotin (ug)	35
Folic acid (ug)	55
B12 (ug)	4.5
C (mg)	20

Minerals

All 22 minerals considered to be essential to the human diet are present in milk. These include three families of salts:

1. *Sodium (Na), Potassium (K) and Chloride (Cl):These* free *ions are negatively correlated to lactose to maintain osmotic equilibrium of milk with blood.*
2. *Calcium (Ca), Magnesium (Mg), Inorganic Phosphorous (P(i)), and Citrate: This group consists of 2/3 of the Ca, 1/3 of the Mg, 1/2 of the P(i), and less than 1/10 of the citrate in* colloidal *(nondiffusible) form and present in the casein micelle.*
3. Diffusible *salts of Ca, Mg, citrate, and phosphate: These salts are very pH dependent and contribute to the overall acid-base equilibrium of milk.*

The mineral content of fresh milk is given below. A wide range is shown, based on variability due to intrinsic (lactogenesis in the secretory cell of individual animals) and extrinsic factors (mainly feed and season/ climate adaptation, but also contaminants). This information is consolidated from C. D. Hunt and F. H. Nielsen, P.L.H. and P.F. Fox, 2009. Advanced Dairy Chemistry, Vol. 3, Lactose, water, salts and minor constituents, 3rd edn.,

Mineral	*Content per litre*
Sodium (mg)	250-640
Potassium (mg)	1100-1500
Chloride (mg)	800-1200
Calcium (mg)	1100-1300
Magnesium (mg)	70-140
Phosphorus (mg)	800-1000
Iron (ug)	100-700
Zinc (ug)	2500-7000
Copper (ug)	100-350
Manganese (ug)	10-50
Iodine (ug)	50-600
Fluoride (ug)	20-80
Selenium (ug)	20-40
Cobalt (ug)	0.5-1.3
Chromium (ug)	0.5-20
Molybdenum (ug)	20-100
Nickel (ug)	0-50
Arsenic (ug)	20-60
Aluminum (ug)	50-1600
Lead (ug)	tr-20

Springer, pp. 392-8; and from, Cristina Sola-Larranaga; Inigo Navarro-Blasco. 2009. Chemometric analysis of minerals and trace elements in raw cow milk from the community of Navarra, Spain. Food Chemistry, 112 (1), 189-196.

Physical Properties

Density

The density of milk and milk products is used for the following;

- to convert volume into mass and vice versa
- to estimate the solids content
- to calculate other physical properties (e.g. kinematic viscosity).

Density, the mass of a certain quantity of material divided by its volume, is dependant on the following:

- temperature at the time of measurement
- temperature history of the material
- composition of the material (especially the fat content)
- inclusion of air (a complication with more viscous products).

With all of this in mind, the density of milk varies within the range of 1027 to 1033 kg m(-3) at 20° C.

Viscosity

Viscosity of milk and milk products is important in determining the following:

- the rate of creaming
- rates of mass and heat transfer
- the flow conditions in dairy processes.

Milk and skim milk, excepting cooled raw milk, exhibit Newtonian behaviour, in which the viscosity is independent of the rate of shear. The viscosity of these products depends on the following:

- Temperature:
  - cooler temperatures increase viscosity due to the increased voluminosity of casein micelles
  - temperatures above 65° C increase viscosity due to the denaturation of whey proteins
- pH: an increase or decrease in pH of milk also causes an increase in casein micelle voluminosity

Cooled raw milk and cream exhibit non-Newtonian behaviour in which the viscosity is dependant on the shear rate. Agitation may cause partial coalescence of the fat globules (partial churning) which increases viscocity. Fat globules that have under gone cold agglutination, may be dispersed due to agitation, causing a decrease in viscosity.

Freezing Point

Freezing point depression is a colligative property which is determined by the molarity of solutes rather than by the percentage by weight or volume. In the dairy industry, freezing point of milk is mainly used to determine added water but it can also been used to determine lactose content in milk, estimate whey powder contents in skim milk powder, and to determine water activity of cheese. The freezing point of milk is usually in the range of-0.512 to-0.550° C with an average of about-0.522° C.

Correct interpretation of freezing point data with respect to added water depends on a good understanding of the factors affecting freezing point depression. With respect to interpretation of freezing points for added water determination, the most significant variables are the nutritional status of the herd and the access to water. Under feeding causes increased freezing points. Large temporary increases in freezing point occur after consumption of large amounts of water because milk is iso-osmotic with blood. The primary sources of non-intentional added water in milk are residual rinse water and condensation in the milking system.

Acid-Base Equilibria

Both titratable acidity and pH are used to measure milk acidity. The pH of milk at 25° C normally varies within a relatively narrow range of 6.5 to 6.7. The normal range for titratable acidity of herd milks is 13 to 20 mmol/L. Because of the large inherent variation, the measure of titratable acidity has little practical value except to measure changes in acidity (eg., during lactic fermentation) and even for this purpose, pH is a better measurement.

There are many components in milk which provide a buffering action. The major buffering groups of milk are caseins and phosphate.

Optical Properties

Optical properties provide the basis for many rapid, indirect methods of analysis such as proximate analysis by infrared absorbency or light scattering. Optical properties also determine the appearance of milk and milk products. Light scattering by fat globules and casein micelles causes milk to appear turbid and opaque. Light scattering occurs when the wave length of light is near the same magnitude as the particle. Thus, smaller particles scatter light of shorter wavelengths. Skim milk appears slightly blue because casein micelles scatter the shorter wavelengths of visible light (blue) more than the red. The carotenoid precursor of vitamin A, ß-carotene, contained in milk fat, is responsible for the 'creamy' colour of milk. Riboflavin imparts a greenish colour to whey.

Refractive index (RI) is normally determined at 20° C with the D line of the sodium spectrum. The refractive index of milk is 1.3440 to 1.3485 and can be used to estimate total solids.

Chapter 3

Microbiology of Dairy Products

Microbiology is the study of *microorganisms*, which are microscopic, unicellular, and cell-cluster organisms. This includes eukaryotes such as fungi and protists, and prokaryotes. Viruses and prions, though not strictly classed as living organisms, are also studied. Microbiology typically includes the study of the immune system, or Immunology. Generally, immune systems interact with pathogenic microbes; these two disciplines often intersect which is why many colleges offer a paired degree such as "Microbiology and Immunology".

Microbiology is a broad term which includes virology, mycology, parasitology, bacteriology and other branches. A microbiologist is a specialist in microbiology and these other topics.

Microbiology is researched actively, and the field is advancing continually. It is estimated only about one percent of all of the microbe species on Earth have been studied. Although microbes were directly observed over three hundred years ago, the field of microbiology can be said to be in its infancy relative to older biological disciplines such as zoology and botany.

History

Ancient

The existence of microorganisms was hypothesized for many centuries before their actual discovery. The existence of unseen microbiological life was postulated by Jainism which is based on Mahavira's teachings as early as 6th century BCE.. Paul Dundas notes that Mahavira asserted existence of unseen microbiological creatures living in earth, water, air and fire. Jain scriptures also describe nigodas which are sub-microscopic creatures living in large clusters and having a very short life and are said to pervade each and every part of the universe, even in tissues of plants and flesh of animals. The Roman Marcus Terentius Varro made references to microbes when he warned against locating a homestead in the vicinity of swamps "because there are bred certain minute creatures which cannot be seen by the eyes, which float in the air and enter the body through the mouth and nose and there cause serious diseases."

In 1546 Girolamo Fracastoro proposed that epidemic diseases were caused by transferable seedlike entities that could transmit infection by direct or indirect contact, or even without contact over long distances.

However, early claims about the existence of microorganisms were speculative, and not based on any data or observation. Actual observation and discovery of microbes had to await the invention of the microscope in the 17th century.

Modern

In 1676, Antonie van Leeuwenhoek observed bacteria and other microorganisms, using a single-lens microscope of his own design. While Van Leeuwenhoek is often cited as the first to observe microbes, Robert Hooke made the first recorded microscopic observation, of the fruiting bodies of molds, in 1665.. The first observation of microbes using a microscope is generally credited to the Dutch draper and haberdasher, Antonie van Leeuwenhoek, who lived for most of his life in Delft, Holland. It has, however, been suggested that a Jesuit priest called Athanasius Kircher was the first to observe microorganisms.

He was among the first to design magic lanterns for projection purposes, so he must have been well acquainted with the properties of lenses. One of his book contains a chapter in Latin, which reads in translation – 'Concerning the wonderful structure of things in nature, investigated by Microscope. Here, he wrote 'who would believe that vinegar and milk abound with an innumerable multitude of worms.' He also noted that putrid material is full of innumerable creeping animalculae. These observations antedate Robert Hooke's Micrographia by nearly 20 years and were published some 29 years before van Leeuwenhoek saw protozoa and 37 years before he described having seen bacteria.

The field of bacteriology (later a subdiscipline of microbiology) was founded in the 19th century by Ferdinand Cohn, a botanist whose studies on algae and photosynthetic bacteria led him to describe several bacteria including *Bacillus* and *Beggiatoa*. Cohn was also the first to formulate a scheme for the taxonomic classification of bacteria and discover spores. Louis Pasteur and Robert Koch were contemporaries of Cohn's and are often considered to be the father of Microbiology and medical microbiology, respectively.

Pasteur is most famous for his series of experiments designed to disprove the then widely held theory of spontaneous generation, thereby solidifying microbiology's identity as a biological science. Pasteur also designed methods for food preservation (pasteurization) and vaccines against several diseases such as anthrax, fowl cholera and rabies.

Koch is best known for his contributions to the germ theory of disease, proving that specific diseases were caused by specific pathogenic microorganisms. He developed a series of criteria that have become known as the Koch's postulates. Koch was one of the first scientists to focus on the isolation of bacteria in pure culture resulting in his description of several novel bacteria including *Mycobacterium tuberculosis*, the causative agent of tuberculosis.

While Pasteur and Koch are often considered the founders of microbiology, their work did not accurately reflect the true diversity of the microbial world because of their exclusive focus on microorganisms having direct medical relevance. It was not until the late 19th century and the work of Martinus Beijerinck and Sergei Winogradsky, the founders of general microbiology (an older term encompassing aspects of microbial physiology, diversity and ecology), that the true breadth of microbiology was revealed. Beijerinck made two major contributions to microbiology: the discovery of viruses and the development of enrichment culture techniques.

While his work on the Tobacco Mosaic Virus established the basic principles of virology, it was his development of enrichment culturing that had the most immediate impact on microbiology by allowing for the cultivation of a wide range of microbes with wildly different physiologies. Winogradsky was the first to develop the concept of chemolithotrophy and to thereby reveal the essential role played by microorganisms in geochemical processes. He was responsible for the first isolation and description of both nitrifying and nitrogen-fixing bacteria.

Fields

The field of microbiology can be generally divided into several subdisciplines:

- *Microbial physiology:* The study of how the microbial cell functions biochemically. Includes the study of microbial growth, microbial metabolism and microbial cell structure.

- *Microbial genetics:* The study of how genes are organized and regulated in microbes in relation to their cellular functions. Closely related to the field of molecular biology.
- *Cellular microbiology:* A discipline bridging microbiology and cell biology.
- *Medical microbiology:* The study of the pathogenic microbes and the role of microbes in human illness. Includes the study of microbial pathogenesis and epidemiology and is related to the study of disease pathology and immunology.
- *Veterinary microbiology:* The study of the role in microbes in veterinary medicine or animal taxonomy.
- *Environmental microbiology:* The study of the function and diversity of microbes in their natural environments. Includes the study of microbial ecology, microbially-mediated nutrient cycling, geomicrobiology, microbial diversity and bioremediation. Characterization of key bacterial habitats such as the rhizosphere and phyllosphere, soil and ground-water ecosystems, open oceans or extreme environments (extremophiles).
- *Evolutionary microbiology:* The study of the evolution of microbes. Includes the study of bacterial systematics and taxonomy.
- *Industrial microbiology:* The exploitation of microbes for use in industrial processes. Examples include industrial fermentation and wastewater treatment. Closely linked to the biotechnology industry. This field also includes brewing, an important application of microbiology.
- *Aeromicrobiology:* The study of airborne microorganisms.
- *Food microbiology:* The study of microorganisms causing food spoilage and foodborne illness. Using microorganisms to produce foods, for example by fermentation.
- *Pharmaceutical microbiology:* the study of microorganisms causing pharmaceutical contamination and spoil
- *Agricultural microbiology:* The study of agriculturally important microorganisms.

(Jobs with the Centre For Disease Control and Prevention requires a degree in microbiology for most positions);

- *Soil Microbiology:* The study of those microorganisms that are found in soil.
- *Water microbiology:* The study of those microorganisms that are found in water.

- *Generation microbiology:* The study of those microorganisms that have the same characters as their parents.
- Nano microbiology: The study of those microorganisms at nano level.

Benefits

Whilst there are undoubtedly some who fear all microbes due to the association of some microbes with various human illnesses, many microbes are also responsible for numerous beneficial processes such as industrial fermentation (e.g. the production of alcohol, vinegar and dairy products), antibiotic production and as vehicles for cloning in more complex organisms such as plants. Scientists have also exploited their knowledge of microbes to produce biotechnological important enzymes such as Taq polymerase, reporter genes for use in other genetic systems and novel molecular biology techniques such as the yeast two-hybrid system.

Bacteria can be used for the industrial production of amino acids. *Corynebacterium glutamicum* is one of the most important bacterial species with an annual production of more than two million tons of amino acids, mainly L-glutamate and L-lysine. A variety of biopolymers, such as polysaccharides, polyesters, and polyamides, are produced by microorganisms. Microorganisms are used for the biotechnological production of biopolymers with tailored properties suitable for high-value medical application such as tissue engineering and drug delivery. Microorganisms are used for the biosynthesis of xanthan, alginate, cellulose, cyanophycin, poly (gamma-glutamic acid), levan, hyaluronic acid, organic acids, oligosaccharides and polysaccharide, and polyhydroxyalkanoates.

Microorganisms are beneficial for microbial biodegradation or bioremediation of domestic, agricultural and industrial wastes and subsurface pollution in soils, sediments and marine environments. The ability of each microorganism to degrade toxic waste depends on the nature of each contaminant. Since sites typically have multiple pollutant types, the most effective approach to microbial biodegradation is to use a mixture of bacterial species and strains, each specific to the biodegradation of one or more types of contaminants.

There are also various claims concerning the contributions to human and animal health by consuming probiotics (bacteria potentially beneficial to the digestive system) and/or prebiotics (substances consumed to promote the growth of probiotic microorganisms).

Recent research has suggested that microorganisms could be useful in the treatment of cancer. Various strains of non-pathogenic clostridia can infiltrate and replicate within solid tumors. Clostridial vectors can be safely administered and their potential to deliver therapeutic proteins has been demonstrated in a variety of preclinical models.

Dairy

A dairy is a building used for the harvesting of animal milk—mostly from cows or goats, but also from buffalo, sheep, horses or camels —for human consumption. A dairy is typically located on a dedicated *dairy farm* or section of a multi-purpose farm that is concerned with the harvesting of milk.

Terminology differs between countries. For example, in the United States, a farm building where milk is harvesting is often called a *milking parlor*. In New Zealand such a building is historically know as the *milking shed* - although in recent years there has been a progressive change to call such a building a *farm dairy*.

In some countries, especially those with small numbers of animals being milked, as well as harvesting the milk from an animal, the dairy may also process the milk into butter, cheese and yoghurt, for example. This is a traditional method of producing specialist milk products, especially in Europe. In the United States a *dairy* can also be a place that processes, distributes and sells dairy products, or a room, building or establishment where milk is stored and processed into milk products, such as butter or cheese. In New Zealand English the singular use of the word *dairy* almost exclusively refers to the corner convenience store, or superette. This usage is historical as such stores were a common place for the public to buy milk products. As an attributive, the word *dairy* refers to milk-based products, derivatives and processes, and the animals and workers involved in their production: for example dairy cattle, dairy goat. A dairy farm produces milk and a dairy factory processes it into a variety of dairy products. These establishments constitute the dairy industry, a component of the food industry.

History

Milk producing animals have been domesticated for thousands of years. Initially, they were part of the subsistence farming that nomads engaged in. As the community moved about the country, their animals accompanied them. Protecting and feeding the animals were a big part of the symbiotic relationship between the animals and the herders.

In the more recent past, people in agricultural societies owned dairy animals that they milked for domestic and local (village) consumption, a typical example of a cottage industry. The animals might serve multiple purposes (for example, as a draught animal for pulling a plough as a youngster, and at the end of its useful life as meat). In this case the animals were normally milked by hand and the herd size was quite small, so that all of the animals could be milked in less than an hour—about 10 per milker. These tasks were performed by a *dairymaid* (*dairywoman*) or *dairyman*. The word *dairy* harkens back to Middle English *dayerie*, *deyerie*, from *deye* (female servant or dairymaid) and further back to Old English *dæge* (kneader of bread).

With industrialisation and urbanisation, the supply of milk became a commercial industry, with specialised breeds of cattle being developed for dairy, as distinct from beef or draught animals. Initially, more people were employed as milkers, but it soon turned to mechanisation with machines designed to do the milking.

Historically, the milking and the processing took place close together in space and time: on a dairy farm. People milked the animals by hand; on farms where only small numbers are kept, hand-milking may still be practiced. Hand-milking is accomplished by grasping the teats (often pronounced *tit* or *tits*) in the hand and expressing milk either by squeezing the fingers progressively, from the udder end to the tip, or by squeezing the teat between thumb and index finger, then moving the hand downward from udder towards the end of the teat. The action of the hand or fingers is designed to close off the milk duct at the udder (upper) end and, by the movement of the fingers, close the duct progressively to the tip to express the trapped milk. Each half or quarter of the udder is emptied one milk-duct capacity at a time.

The *stripping* action is repeated, using both hands for speed. Both methods result in the milk that was trapped in the milk duct being squirted out the end into a bucket that is supported between the knees (or rests on the ground) of the milker, who usually sits on a low stool. Traditionally the cow, or cows, would stand in the field or paddock while being milked. Young stock, heifers, would have to be trained to remain still to be milked. In many countries, the cows were tethered to a post and milked. The problem with this method is that it relies on quiet, tractable beasts, because the hind end of the cow is not restrained.

In 1937, it was found that bovine somatotropin (bST or bovine growth hormone) would increase the yield of milk. Monsanto Company developed a synthetic (recombinant) version of this hormone (rBST). In February 1994, rBST was approved by the Food and Drug Administration (FDA) for use in the U.S. It has become common in the U.S., but not elsewhere, to inject it into milch kine dairy cows to increase their production by up to 15%.

However, there are claims that this practice can have negative consequences for the animals themselves. A European Union scientific commission was asked to report on the incidence of mastitis and other disorders in dairy cows, and on other aspects of the welfare of dairy cows. The commission's statement, subsequently adopted by the European Union, stated that the use of rBST substantially increased health problems with cows, including foot problems, mastitis and injection site reactions, impinged on the welfare of the animals and caused reproductive disorders. The report concluded that on the basis of the health and welfare of the animals, rBST should not be used. Health Canada prohibited the sale of rBST in 1999; the recommendations of external committees were that, despite not finding a significant health risk to humans, the drug presented a threat to animal health and, for this reason, could not be sold in Canada.

Structure of the Industry

While most countries produce their own milk products, the structure of the dairy industry varies in different parts of the world. In major milk-producing countries most milk is distributed through wholesale markets. In Ireland and Australia, for example, farmers' co-operatives own many of the large-scale processors, while in the United States many farmers and processors do business through individual contracts. In the United States, the country's 196 farmers' cooperatives sold 86% of milk in the U.S. in 2002, with five cooperatives accounting for half that. This was down from 2,300 cooperatives in the 1940s. In developing countries, the past practice of farmers marketing milk in their own neighbourhoods are changing rapidly. Notable developments include considerable foreign investment in the dairy industry and a growing role for dairy cooperatives. Output of milk is growing rapidly in such countries and presents a major source of income growth for many farmers.

As in many other branches of the food industry, dairy processing in the major dairy producing countries has become increasingly concentrated, with fewer but larger and more efficient plants operated by fewer workers. This is notably the case in the United States, Europe, Australia and New Zealand. In 2009, charges of anti-trust violations have been made against major dairy industry players in the United States.

Government intervention in milk markets was common in the 20th century. A limited anti-trust exemption was created for U.S. dairy cooperatives by the Capper-Volstead Act of 1922. In the 1930s, some U.S. states adopted price controls, and Federal Milk Marketing Orders started under the Agricultural Marketing Agreement Act of 1937 and continue in the 2000s. The Federal Milk Price Support Program began in 1949. The Northeast Dairy Compact regulated wholesale milk prices in New England from 1997 to 2001.

Plants producing liquid milk and products with short shelf life, such as yogurts, creams and soft cheeses, tend to be located on the outskirts of urban centres close to consumer markets. Plants manufacturing items with longer shelf life, such as butter, milk powders, cheese and whey powders, tend to be situated in rural areas closer to the milk supply. Most large processing plants tend to specialise in a limited range of products. Exceptionally, however, large plants producing a wide range of products are still common in Eastern Europe, a holdover from the former centralized, supply-driven concept of the market.

As processing plants grow fewer and larger, they tend to acquire bigger, more automated and more efficient equipment. While this technological tendency keeps manufacturing costs lower, the need for long-distance transportation often increases the environmental impact.

Milk production is irregular, depending on cow biology. Producers must adjust the mix of milk which is sold in liquid form vs. processed foods (such as butter and cheese) depending on changing supply and demand.

Operation of the Dairy Farm

When it became necessary to milk larger numbers of cows, the cows would be brought to a shed or barn that was set up with bails (stalls) where the cows could be confined while they were milked. One person could milk more cows this way, as many as 20 for a skilled worker. But having cows standing about in the yard and shed waiting to be milked is not good for the cow, as she needs as much time in the paddock grazing as is possible. It is usual to restrict the twice-daily milking to a maximum of an hour and a half each time. It makes no difference whether one milks 10 or 1000 cows, the milking time should not exceed a total of about three hours each day for any cow.

As herd sizes increased there was more need to have efficient milking machines, sheds, milk-storage facilities (vats), bulk-milk transport and shed cleaning capabilities and the means of getting cows from paddock to shed and back.

Farmers found that cows would abandon their grazing area and walk towards the milking area when the time came for milking. This is not surprising as, in the flush of the milking season, cows presumably get very uncomfortable with udders engorged with milk, and the place of relief for them is the milking shed.

As herd numbers increased so did the problems of animal health. In New Zealand two approaches to this problem have been used. The first was improved veterinary medicines (and the government regulation of the medicines) that the farmer could use. The other was the creation of *veterinary clubs* where groups of farmers would employ a veterinarian (vet) full-time and share those services throughout the year. It was in the vet's interest to keep the animals healthy and reduce the number of calls from farmers, rather than to ensure that the farmer needed to call for service and pay regularly. Most dairy farmers milk their cows with absolute regularity at a minimum of twice a day, with some high-producing herds milking up to four times a day to lessen the weight of large volumes of milk in the udder of the cow. This daily milking routine goes on for about 300 to 320 days per year that the cow stays in milk. Some small herds are milked once a day for about the last 20 days of the production cycle but this is not usual for large herds.

If a cow is left unmilked just once she is likely to reduce milk-production almost immediately and the rest of the season may see her *dried off* (giving no milk) and still consuming feed for no production. However, once-a-day milking is now being practised more widely in New Zealand for profit and lifestyle reasons. This is effective because the fall in milk yield is at least partially offset by labour and cost savings from milking once per day. This compares to some intensive farm systems in the United States that milk three or more times per day due to higher milk yields per cow and lower marginal labour costs.

Farmers who are contracted to supply liquid milk for human consumption (as opposed to milk for processing into butter, cheese, and so on—see milk) often have to manage their herd so that the contracted number of cows are in milk the year round, or the required minimum milk output is maintained. This is done by mating cows outside their natural mating time so that the period when each cow in the herd is giving maximum production is in rotation throughout the year.

Northern hemisphere farmers who keep cows in barns almost all the year usually manage their herds to give continuous production of milk so that they get paid all year round. In the southern hemisphere the cooperative dairying systems allow for two months on no productivity because their systems are designed to take advantage of maximum grass and milk production in the spring and because the milk processing plants pay bonuses in the dry (winter) season to carry the farmers through the mid-winter break from milking. It also means that cows have a rest from milk production when they are most heavily pregnant. Some year-round milk farms are penalised financially for over-production at any time in the year by being unable to sell their overproduction at current prices.

Artificial insemination (AI) is common in all high-production herds.

Industrial Processing

Dairy plants process the raw milk they receive from farmers so as to extend its marketable life. Two main types of processes are employed: heat treatment to ensure the safety of milk for human consumption and to lengthen its shelf-life, and dehydrating dairy products such as butter, hard cheese and milk powders so that they can be stored.

Cream and Butter

Today, milk is separated by large machines in bulk into cream and skim milk. The cream is processed to produce various consumer products, depending on its thickness, its suitability for culinary uses and consumer demand, which differs from place to place and country to country. Some cream is dried and powdered, some is condensed (by evaporation) mixed with varying amounts of sugar and canned. Most cream from New Zealand and Australian factories is made into butter. This is done by churning the cream until the fat globules coagulate and form a monolithic mass. This butter mass is washed and, sometimes, salted to improve keeping qualities. The residual buttermilk goes on to further processing. The butter is packaged (25 to 50 kg boxes) and chilled for storage and sale. At a later stage these packages are broken down into home-consumption sized packs.

Skimmed Milk

The product left after the cream is removed is called skim, or skimmed, milk. To make a consumable liquid a portion of cream is returned to the skim milk to make *low fat milk* (semi-skimmed) for human consumption. By varying the amount of cream returned, producers can make a variety of low-fat milks to suit their local market. Other products, such as calcium, vitamin D, and flavouring, are also added to appeal to consumers.

Casein

Casein is the predominant phosphoprotein found in fresh milk. It has a very wide range of uses from being a filler for human foods, such as in ice cream, to the manufacture of products such as fabric, adhesives, and plastics.

Cheese

Cheese is another product made from milk. Whole milk is reacted to form curds that can be compressed, processed and stored to form cheese. In countries where milk is legally allowed to be processed without pasteurisation a wide range of cheeses can be made using the bacteria naturally in the milk. In most other countries, the range of cheeses is smaller and the use of artificial cheese curing is greater. Whey is also the by-product of this process.

Whey

In earlier times whey was considered to be a waste product and it was, mostly, fed to pigs as a convenient means of disposal. Beginning about 1950, and mostly since about 1980, lactose and many other products, mainly food additives, are made from both casein and cheese whey.

Yogurt

Yoghurt (or yogurt) making is a process similar to cheese making, only the process is arrested before the curd becomes very hard.

Milk Powders

Milk is also processed by various drying processes into powders. Whole milk, skim milk, buttermilk, and whey products are dried into a powder form and used for human and animal consumption. The main difference between production of powders for human or for animal consumption is in the protection of the process and the product from contamination. Some people drink milk reconstituted from powdered milk, because milk is about 88% water and it is much cheaper to transport the dried product.

Other Milk Products

Kumis is produced commercially in Central Asia. Although it is traditionally made from mare's milk, modern industrial variants may use cow's milk instead.

Transport of Milk

Historically, the milking and the processing took place in the same place: on a dairy farm. Later, cream was separated from the milk by machine, on the farm, and the cream was transported to a factory for butter making. The skim milk was fed to pigs. This allowed for the high cost of transport (taking the smallest volume high-value product), primitive trucks and the poor quality of roads. Only farms close to factories could afford to take whole milk, which was essential for cheese making in industrial quantities, to them. The development of refrigeration and better road transport, in the late 1950s, has meant that most farmers milk their cows and only temporarily store the milk in large refrigerated bulk tanks, from where it is later transported by truck to central processing facilities.

Milking Machines

Milking machines are used to harvest milk from cows when manual milking becomes inefficient or labour intensive. The milking unit is the portion of a milking machine for removing milk from an udder. It is made up of a claw, four teatcups, (Shells and rubber liners) long milk tube, long pulsation tube, and a pulsator. The claw is an assembly that connects the short pulse tubes and short milk tubes from the teatcups to the long pulse tube and long milk tube. (Cluster assembly) Claws are commonly made of stainless steel or plastic or both. Teatcups are composed of a rigid outer shell (stainless steel or plastic) that holds a soft inner liner or *inflation.* Transparent sections in the shell may allow viewing of liner collapse and milk flow. The annular space between the shell and liner is called the pulse chamber.

Milking machines work in a way that is different from hand milking or calf suckling. Continuous vacuum is applied inside the soft liner to massage milk from the teat by creating a pressure difference across the teat canal (or opening at the end of the teat). Vacuum also helps keep the machine attached to the cow. The vacuum applied to the teat causes congestion of teat tissues (accumulation of blood and other fluids). Atmospheric air is admitted into the pulsation chamber about once per second (the pulsation rate) to allow the liner to collapse around the end of teat and relieve congestion in the teat tissue. The ratio of the time that the liner is open (milking phase) and closed (rest phase) is called the pulsation ratio.

The four streams of milk from the teatcups are usually combined in the claw and transported to the milkline, or the collection bucket (usually sized to the output of one cow) in a single milk hose. Milk is then transported (manually in buckets) or with a combination of airflow and mechanical pump to a central storage vat or bulk tank. Milk is refrigerated on the farm in most countries either by passing through a heat-exchanger or in the bulk tank, or both.

In the photo above is a bucket milking system with the stainless steel bucket visible on the far side of the cow. The two rigid stainless steel teatcup shells applied to the front two quarters of the udder are visible. The top of the flexible liner is visible at the top of the shells as are the short milk tubes and short pulsation tubes extending from the bottom of the shells to the claw. The bottom of the claw is transparent to allow observation of milk flow. When milking is completed the vacuum to the milking unit is shut off and the teatcups are removed.

Milking machines keep the milk enclosed and safe from external contamination. The interior 'milk contact' surfaces of the machine are kept clean by a manual or automated washing procedures implemented after milking is completed. Milk contact surfaces must comply with regulations requiring food-grade materials (typically stainless steel and special plastics and rubber compounds) and are easily cleaned.

Most milking machines are powered by electricity but, in case of electrical failure, there can be an alternative means of motive power, often an internal combustion engine, for the vacuum and milk pumps. Milk cows cannot tolerate delays in scheduled milking without serious milk production reductions.

Milking Shed Layouts

Bail-style sheds— This type of milking facility was the first development, after open-paddock milking, for many farmers. The building was a long, narrow, *lean-to* shed that was open along one long side. The cows were held in a yard at the open side and when they were about to be milked they were positioned in one of the bails (stalls). Usually the cows were restrained in the bail with a breech chain and a rope to restrain the outer back leg. The cow could not move about excessively and the milker could expect not to be kicked or trampled while sitting on a (three-legged) stool and milking into a bucket. When each cow was finished she backed out into the yard again. The UK bail, developed largely by Rex Patterson, was a six standing mobile shed with steps that the cow mounted, so the herdsman didn't have to bend so low. The milking equipment was much as today, a vacuum from a pump, pulsators, a claw-piece with pipes leading to the four shells and liners that stimulate and suck the milk from the teat. The milk went into churns, via a cooler.

As herd sizes increased a door was set into the front of each bail so that when the milking was done for any cow the milker could, after undoing the leg-rope and with a remote link, open the door and allow her to exit to the pasture. The door was closed, the next cow walked into the bail and was secured. When milking machines were introduced bails were set in pairs so that a cow was being milked in one paired bail while the other could be prepared for milking. When one was finished the machine's cups are swapped to the other cow. This is the same as for *Swingover Milking Parlours* as described below except that the cups are loaded on the udder from the side. As herd numbers increased it was easier to double-up the cup-sets and milk both cows simultaneously than to increase the number of bails. About 50 cows an hour can be milked in a shed with 8 bales by one person. Using the same teat cups for successive cows has the danger of transmitting infection, mastitis, from one cow to another. Some farmers have devised their own ways to disinfect the clusters between cows.

Herringbone Milking Parlours— In herringbone milking sheds, or parlours, cows enter, in single file, and line up almost perpendicular to the central aisle of the milking parlour on both sides of a central pit in which the milker works (you can visualise a fishbone with the ribs representing the cows and the spine being the milker's working area; the cows face outward). After washing the udder and teats the cups of the milking machine are applied to the cows, from the rear of their hind legs, on both sides of the working area. Large herringbone sheds can milk up to 600 cows efficiently with two people.

Swingover Milking Parlours— Swingover parlours are the same as herringbone parlours except they have only one set of milking cups to be shared between the two rows of cows, as one side is being milked the cows on the other side are moved out and replaced with unmilked ones. The advantage of this system is that it is less costly to equip, however it operates at slightly better than half-speed and one would not normally try to milk more than about 100 cows with one person.

Rotary Milking sheds— Rotary milking sheds consist of a turntable with about 12 to 100 individual stalls for cows around the outer edge. A "good" rotary will be operated with 24–32 (~48–50+) stalls by one (two) milkers. The turntable is turned by an electric-motor drive at a rate that one turn is the time for a cow to be milked completely. As an empty stall passes the entrance a cow steps on, facing the centre, and rotates with the turntable.

The next cow moves into the next vacant stall and so on. The operator, or milker, cleans the teats, attaches the cups and does any other feeding or whatever husbanding operations that are necessary. Cows are milked as the platform rotates. The milker, or an automatic device, removes the milking machine cups and the cow backs out and leaves at an exit just before the entrance. The rotary system is capable of milking very large herds—over a thousand cows.

Automatic Milking sheds— Automatic milking or 'robotic milking' sheds can be seen in Australia, New Zealand and many European countries. Current automatic milking sheds use the voluntary milking (VM) method. These allow the cows to voluntarily present themselves for milking at any time of the day or night, although repeat visits may be limited by the farmer through computer software. A robot arm is used to clean teats and apply milking equipment, while automated gates direct cow traffic, eliminating the need for the farmer to be present during the process. The entire process is computer controlled.

Supplementary accessories in sheds— Farmers soon realised that a milking shed was a good place to feed cows supplementary foods that overcame local dietary deficiencies or added to the cows' wellbeing and production. Each bail might have a box into which such feed is delivered as the cow arrives so that she is eating while being milked. A computer can read the eartag of each beast to ration the correct individual supplement. A close alternative is to use 'out-of-parlour-feeders', stalls that respond to a transponder around the cow's neck that is programmed to provide each cow with a supplementary feed, the quantity dependent on her production, stage in lactation, and the benefits of the main ration.

The holding yard at the entrance of the shed is important as a means of keeping cows moving into the shed. Most yards have a powered gate that ensures that the cows are kept close to the shed.

Water is a vital commodity on a dairy farm: cows drink about 20 gallons (80 litres) a day, sheds need water to cool and clean them. Pumps and reservoirs are common at milking facilities. Water can be warmed by heat transfer with milk.

Temporary Milk Storage

Milk coming from the cow is transported to a nearby storage vessel by the airflow leaking around the cups on the cow or by a special "air inlet" (5-10 l/min free air) in the claw. From there it is pumped by a mechanical pump and cooled by a heat exchanger. The milk is then stored in a large vat, or bulk tank, which is usually refrigerated until collection for processing.

Processing Facilities

Topics:

- Pasteurization, homogenization
- Cream extraction
- Cheese making
- Butter making
- Caseinmaking
- Yogurt processing.

Waste Disposal

In countries where cows are grazed outside year-round, there is little waste disposal to deal with. The most concentrated waste is at the milking shed, where the animal waste is liquefied (during the water-washing process) and allowed to flow by gravity, or pumped, into composting ponds with anaerobic bacteria to consume the solids. The processed water and nutrients are then pumped back onto the pasture as irrigation and fertilizer. Surplus animals are slaughtered for processed meat and other rendered products.

In the associated milk processing factories, most of the waste is washing water that is treated, usually by composting, and returned to waterways. This is much different from half a century ago, when the main products were butter, cheese and casein, and the rest of the milk had to be disposed of as waste (sometimes as animal feed).

In areas where cows are housed all year round, the waste problem is difficult because of the amount of feed that is brought in and the amount of bedding material that also has to be removed and composted. The size of the problem can be understood by standing downwind of the barns where such dairying goes on.

In many cases, modern farms have very large quantities of milk to be transported to a factory for processing. If anything goes wrong with the milking, transport or processing facilities it can be a major disaster trying to dispose of enormous quantities of milk. If a road tanker overturns on a road, the rescue crew is looking at accommodating the spill of 5 to 10 thousand gallons of milk (20 to 45 thousand litres) without allowing any into the waterways.

A derailed rail tanker-train may involve 10 times that amount. Without refrigeration, milk is a fragile commodity, and it is very damaging to the environment in its raw state due to its high biochemical oxygen demand. A widespread electrical power blackout is another disaster for the dairy industry, because both milking and processing facilities are affected. For this, farms may often use mobile generators. Such a situation occurred during the power outage caused by the 2010 Canterbury Earthquake.

In dairy-intensive areas, various methods have been proposed for disposing of large quantities of milk. These directives include feeding milk to livestock, spray irrigation or designating a sacrifice area. Large application rates of milk onto land, or disposing in a hole, is problematic as the residue from the decomposing milk will block the soil pores and thereby reduce the water infiltration rate through the soil profile. As recovery of this effect can take time, any land based application needs to be well managed and considered.

Associated Diseases

- Leptospirosis is one of the most common debilitating diseases of milkers, made somewhat worse since the introduction of herringbone sheds, because of unavoidable direct contact with bovine urine

- Cowpox is one of the helpful diseases; it is barely harmful to humans and tends to inoculate them against other poxes such as small pox.
- Tuberculosis (TB) is able to be transmitted from cattle mainly via milk products that are unpasteurised. TB has been eradicated from many countries by testing for the disease and culling suspected animals.
- Brucellosis is a bacterial disease transmitted to humans by dairy products and direct animal contact. Brucellosis has been eradicated from certain countries by testing for the disease and culling suspected animals
- Listeria is a bacterial disease associated with unpasteurised milk, and can affect some cheeses made in traditional ways. Careful observance of the traditional cheese making methods achieves reasonable protection for the consumer.
- Johne's Disease (pronounced "yo-knees") is a contagious, chronic and sometimes fatal infection in ruminants caused by a bacterium named Mycobacterium avium subspecies paratuberculosis (M. paratuberculosis). The bacteria are present in retail milk, and are believed by some researchers to be the primary cause of Crohn's disease in humans. This disease is not known to infect animals in Australia and New Zealand.

Microorganisms

Microorganisms are living organisms that are individually too small to see with the naked eye. The unit of measurement used for microorganisms is the micrometer (μ m); 1 μ m = 0.001 millimeter; 1 nanometer (nm) = 0.001 μ m. Microorganisms are found everywhere (ubiquitous) and are essential to many of our planets life processes. With regards to the food industry, they can cause spoilage, prevent spoilage through fermentation, or can be the cause of human illness.

There are several classes of microorganisms, of which bacteria and fungi (yeasts and moulds) will be discussed in some detail. Another type of microorganism, the bacterial viruses or bacteriophage, will be examined in a later section.

Bacteria

Bacteria are relatively simple single-celled organisms. One method of classification is by shape or morphology:

- Cocci:
  - — spherical shape
  - — 0.4 - 1.5 μ m

Examples: staphylococci - form grape-like clusters; streptococci - form bead-like chains;

- Rods:
  - 0.25 - 1.0 μ m width by 0.5 - 6.0 μ m long.

Examples: bacilli - straight rod; spirilla - spiral rod.

There exists a bacterial system of taxonomy, or classification system, that is internationally recognized with family, genera and species divisions based on genetics.

Some bacteria have the ability to form resting cells known as endospores. The spore forms in times of environmental stress, such as lack of nutrients and moisture needed for growth, and thus is a survival strategy. Spores have no metabolism and can withstand adverse conditions such as heat, disinfectants, and ultraviolet light. When the environment becomes favourable, the spore germinates and giving rise to a single vegetative bacterial cell. Some examples of spore-formers important to the food industry are members of *Bacillus* and *Clostridium* generas.

Bacteria reproduce asexually by fission or simple division of the cell and its contents. The doubling time, or generation time, can be as short as 20-20 min. Since each cell grows and divides at the same rate as the parent cell, this could under favourable conditions translate to an increase from one to 10 million cells in 11 hours! However, bacterial growth in reality is limited by lack of nutrients, accumulation of toxins and metabolic wastes, unfavourable temperatures and dessication. The maximum number of bacteria is approximately 1 X 10e9 CFU/g or ml.

Note: Bacterial populations are expressed as colony forming units (CFU) per gram or millilitre.

Bacterial growth generally proceeds through a series of phases:

- Lag phase: time for microorganisms to become accustomed to their new environment. There is little or no growth during this phase.
- Log phase: bacteria logarithmic, or exponential, growth begins; the rate of multiplication is the most rapid and constant.
- Stationary phase: the rate of multiplication slows down due to lack of nutrients and build-up of toxins. At the same time, bacteria are constantly dying so the numbers actually remain constant.
- Death phase: cell numbers decrease as growth stops and existing cells die off.

The shape of the curve varies with temperature, nutrient supply, and other growth factors. This exponential death curve is also used in modeling the heating destruction of microorganisms.

Yeasts

Yeasts are members of a higher group of microorganisms called fungi . They are single-cell organisms of spherical, elliptical or cylindrical shape. Their size varies greatly but are generally larger than bacterial cells. Yeasts may be divided into two groups according to their method of reproduction:

1. budding: called Fungi Imperfecti or false yeasts
2. budding and spore formation: called Ascomycetes or true yeasts.

Unlike bacterial spores, yeast form spores as a method of reproduction.

Moulds

Moulds are filamentous, multi-celled fungi with an average size larger than both bacteria and yeasts (10 X 40 μ m). Each filament is referred to as a hypha. The mass of hyphae that can quickly spread over a food substrate is called the mycelium. Moulds may reproduce either asexually or sexually, sometimes both within the same species.

Asexual Reproduction:

- fragmentation - hyphae separate into individual cells called arthropsores
- spore production - formed in the tip of a fruiting hyphae, called conidia, or in swollen structures called sporangium.

Sexual Reproduction: sexual spores are produced by nuclear fission in times of unfavourable conditions to ensure survival.

Microbial Growth

There are a number of factors that affect the survival and growth of microorganisms in food. The parameters that are inherent to the food, or intrinsic factors, include the following:

- nutrient content
- moisture content
- pH
- available oxygen
- biological structures
- antimicrobial constituents.

Nutrient Requirements: While the nutrient requirements are quite organism specific, the microorganisms of importance in foods require the following:

- water
- energy source
- carbon/nitrogen source
- vitamins
- minerals.

Milk and dairy products are generally very rich in nutrients which provides an ideal growth environment for many microorganisms.

Moisture Content: All microorganisms require water but the amount necessary for growth varies between species. The amount of water that is available in food is expressed in terms of water activity (aw), where the aw of pure water is 1.0. Each microorganism has a maximum, optimum, and minimum aw for growth and survival. Generally bacteria dominate in foods with high aw (minimum approximately 0.90 aw) while yeasts and moulds, which require less moisture, dominate in low aw foods (minimum 0.70 aw). The water activity of fluid milk is approximately 0.98 aw.

pH: Most microorganisms have approximately a neutral pH optimum (pH 6-7.5). Yeasts are able to grow in a more acid environment compared to bacteria. Moulds can grow over a wide pH range but prefer only slightly acid conditions. Milk has a pH of 6.6 which is ideal for the growth of many microoorganisms.

Available Oxygen: Microorganisms can be classified according to their oxygen requirements necessary for growth and survival:

- Obligate Aerobes: oxygen required
- Facultative: grow in the presence or absence of oxygen
- Microaerophilic: grow best at very low levels of oxygen
- Aerotolerant Anaerobes: oxygen not required for growth but not harmful if present
- Obligate Anaerobes: grow only in complete absence of oxygen; if present it can be lethal.

Biological Structures: Physical barriers such as skin, rinds, feathers, etc. have provided protection to plants and animals against the invasion of microorganisms. Milk, however, is a fluid product with no barriers to the spreading of microorganisms throughout the product.

Antimicrobial Constituents: As part of the natural protection against microorganisms, many foods have antimicrobial factors. Milk has several nonimmunological proteins which inhibit the growth and metabolism of many microorganisms including the following most common:

1. lactoperoxidase
2. lactoferrin
3. lysozyme
4. xanthine.

Where the intrinsic factors are related to the food properties, the extrinsic factors are related to the storage environment. These would include temperature, relative humidity, and gases that surround the food.

Temperature: As a group, microorganisms are capable of growth over an extremely wide temperature range. However, in any particular environment, the types and numbers of microorganisms will depend greatly on the temperature. According to temperature, microorganisms can be placed into one of three broad groups:

- Psychrotrophs: optimum growth temperatures 20 to 30° capable of growth at temperatures less than 7° C. Psychrotrophic organisms are specifically important in the spoilage of refrigerated dairy products.
- Mesophiles: optimum growth temperatures 30 to 40° C; do not grow at refrigeration temperatures
- Thermophiles: optimum growth between 55 and 65° C

It is important to note that for each group, the growth rate increases as the temperature increases only up to an optimum, afterwhich it rapidly declines.

Detection and Enumeration of Microorganisms

There are several methods for detection and enumeration of microorganisms in food. The method that is used depends on the purpose of the testing.

Direct Enumeration

Using direct microscopic counts (DMC), Coulter counter etc. allows a rapid estimation of all viable and nonviable cells. Identification through staining and observation of morphology also possible with DMC.

Viable Enumeration

The use of standard plate counts, most probable number (MPN), membrane filtration, plate loop methos, spiral plating etc., allows the estimation of only viable cells. As with direct enumeration, these methods can be used in the food industry to enumerate fermentation, spoilage, pathogenic, and indicator organisms.

Metabolic Activity Measurement

An estimation of metabolic activity of the total cell population is possible using dye reduction tests such as resazurin or methylene blue dye reduction, acid production, electrical impedence etc. The level of bacterial activity can be used to assess the keeping quality and freshness of milk. Toxin levels can also be measured, indicating the presence of toxin producing pathogens.

Cellular Constituents Measurement

Using the luciferase test to measure ATP is one example of the rapid and sensitive tests available that will indicate the presence of even one pathogenic bacterial cell.

Isolation of microorganisms is an important preliminary step in the identification of most food spoilage and pathogenic organisms. This can be done using a simple streak plate method.

Microorganisms in Milk

Milk is sterile at secretion in the udder but is contaminated by bacteria even before it leaves the udder. Except in the case of mastisis, the bacteria at this point are harmless and few in number. Further infection of the milk by microorganisms can take place during milking, handling, storage, and other pre-processing activities.

Lactic acid bacteria: this group of bacteria are able to ferment lactose to lactic acid. They are normally present in the milk and are also used as starter cultures in the production of cultured dairy products such as yogurt. Note: many lactic acid bacteria have recently been reclassified; the older names will appear in brackets as you will still find the older names used for convenience sake in a lot of literature. Some examples in milk are:

- lactococci
  - *L. delbrueckii* subsp. *lactis* (*Streptococcus lactis*)
  - *Lactococcus lactis* subsp. *cremoris* (*Streptococcus cremoris*)
- lactobacilli
  - *Lactobacillus casei*
  - *L.delbrueckii* subsp. *lactis* (*L. lactis*)
  - *L. delbrueckii* subsp. *bulgaricus* (*Lactobacillus bulgaricus*)
- *Leuconostoc.*

Coliforms: coliforms are facultative anaerobes with an optimum growth at 37° C. Coliforms are indicator organisms; they are closely associated with the presence of pathogens but not necessarily pathogenic themselves.

They also can cause rapid spoilage of milk because they are able to ferment lactose with the production of acid and gas, and are able to degrade milk proteins. They are killed by HTST treatment, therefore, their presence after treatment is indicative of contamination. *Escherichia coli* is an example belonging to this group.

Significance of microorganisms in milk:

- Information on the microbial content of milk can be used to judge its sanitary quality and the conditions of production
- If permitted to multiply, bacteria in milk can cause spoilage of the product
- Milk is potentially susceptible to contamination with pathogenic microorganisms. Precautions must be taken to minimize this possibility and to destroy pathogens that may gain entrance
- Certain microorganisms produce chemical changes that are desirable in the production of dairy products such as cheese, yogurt.

Spoilage Microorganisms in Milk

The microbial quality of raw milk is crucial for the production of quality dairy foods. Spoilage is a term used to describe the deterioration of a foods' texture, colour, odour or flavour to the point where it is unappetizing or unsuitable for human consumption. Microbial spoilage of food often involves the degradation of protein, carbohydrates, and fats by the microorganisms or their enzymes.

In milk, the microorganisms that are principally involved in spoilage are psychrotrophic organisms. Most psychrotrophs are destroyed by pasteurization temperatures, however, some like *Pseudomonas fluorescens, Pseudomonas fragi* can produce proteolytic and lipolytic extracellular enzymes which are heat stable and capable of causing spoilage.

Some species and strains of *Bacillus, Clostridium, Cornebacterium, Arthrobacter, Lactobacillus, Microbacterium, Micrococcus,* and *Streptococcus* can survive pasteurization and grow at refrigeration temperatures which can cause spoilage problems.

Pathogenic Microorganisms in Milk

Hygienic milk production practices, proper handling and storage of milk, and mandatory pasteurization has decreased the threat of milkborne diseases such as tuberculosis, brucellosis, and typhoid fever. There have been a number of foodborne illnesses resulting from the ingestion of raw milk, or dairy products made with milk that was not properly pasteurized or was poorly handled causing post-processing contamination. The following bacterial pathogens are still of concern today in raw milk and other dairy products:

- *Bacillus cereus*
- *Listeria monocytogenes*
- *Yersinia enterocolitica*
- *Salmonella* spp.
- *Escherichia coli* O157:H7
- *Campylobacter jejuni* .

It should also be noted that moulds, mainly of species of *Aspergillus, Fusarium,* and *Penicillium* can grow in milk and dairy products. If the conditions permit, these moulds may produce mycotoxins which can be a health hazard.

HACCP

Raw and end-products may be tested for the presence, level, or absence of microorganisms. Traditionally these practices were used to reduce manufacturing defects in dairy products and ensure compliance with specifications and regulations, however, they have many drawbacks:

1. destructive and time consuming
2. slow response
3. small sample size
4. delays in the release of the food.

In the 1960's, the Pillsbury Company, the U.S. Army, and NASA introduced a system for assuring pathogen-free foods for the space program. This system, called Hazard Analysis and Critical Control Points (HACCP), is a focus on critical food safety areas as part of total quality programs. It involves a critical examination of the entire food manufacturing process to determine every step where there is a possibility of physical, chemical, or microbiological contamination of the food which would render it unsafe or unacceptable for human consumption. These identified points are the critical control points (CCP). There are seven prinicples to HACCP:

1. analyze hazards
2. determine CCPs

3. establish critical limits
4. establish monitoring procedures
5. establish deviation procedures
6. establish verification procedures
7. establish record keeping procedures.

Before these principles can be put into place, a prerequisite program and preliminary setup is necessary.

Prerequisite Program:

- premise control
- receiving and storage control
- equipment performance and maintenance control
- personnel training
- sanitation
- recall procedure.

Preliminary Setup:

- assemble team
- describe the product
- identify intended use
- construct flow diagram and plant schematic
- verify the diagram on-site.

Food Safety Enhancement Program-FSEP is The Canadian Food Inspection Agency's HACCP initiative. There is extensive information at their Web site regarding FSEP, including implementation manuals, HACCP curriculum guidelines, and generic models.

Starter Cultures

Starter cultures are those microorganisms that are used in the production of cultured dairy products such as yogurt and cheese. The natural microflora of the milk is either inefficient, uncontrollable, and unpredictable, or is destroyed altogether by the heat treatments given to the milk. A starter culture can provide particular characteristics in a more controlled and predictable fermentation. The primary function of lactic starters is the production of lactic acid from lactose. Other functions of starter cultures may include the following:

- flavour, aroma, and alcohol production
- proteolytic and lipolytic activities
- inhibition of undesirable organisms.

There are two groups of lactic starter cultures:

1. simple or defined: single strain, or more than one in which the number is known
2. mixed or compound: more than one strain each providing its own specific characteristics.

Starter cultures may be categorized as mesophilic or thermophilic:

Mesophilic

- *Lactococcus lactis* subsp. *cremoris*
- *L. delbrueckii* subsp. *lactis*
- *L. lactis* subsp. *lactis* biovar *diacetylactis*
- *Leuconostoc mesenteroides* subsp. *cremoris.*

Thermophilic

- *Streptococcus salivarius* subsp. *thermophilus* (*S.thermophilus*)
- *Lactobacillus delbrueckii* subsp. *bulgaricus*
- *L. delbrueckii* subsp. *lactis*
- *L. casei*
- *L. helveticus*
- *L. plantarum* .

Mixtures of mesophilic and thermophilic microorganisms can also be used as in the production of some cheeses.

Bacteriophage

Bacteriophages are viruses that require bacteria host cells for growth and reproduction. Initially, the bacteriophage attaches itself to the bacteria cell wall and injects nuclear substance into the cell. Inside the cell, the nuclear substance produces shells, or phage coats, for the new bacteriophage which are quickly filled with nucleic acid. The bacterial cell ruptures and dies as the new bacteriophage are released. Bacteriophages are ubiquitous but generally enter the milk processing plant with the farm milk. They can be inactivated heat treatments of 30 min at 63 to 88° C, or by the use of chemical disinfectants.

Bacteriophages are of most concern in cheese making. They attack and destroy most of the lactic acid bacteria which prevents normal ripening known as slow or dead vat.

Starter Culture Preparation

Commercial manufacturers provide starter cultures in lyophilized (freeze-dryed), frozen or spray-dried forms. The dairy product manufacturers need to inoculate the culture into milk or other suitable substrate. There are a number of steps necessary for the propagation of starter culture ready for production:

1. Commercial culture
2. Mother culture — first inoculation; all cultures will originate from this preparation
3. Intermediate culture — in preparation of larger volumes of prepared starter
4. Bulk starter culture — this stage is used in dairy product production.

Chapter 4

Physico-chemical Testing of Milk and Dairy Products

Microbiological and physicochemical analysis of different UHT milks available in market Raw milk is milk in its natural (unpasteurized) state. Contaminated raw milk can be a source of harmful bacteria, such as those that cause undulant fever, dysentery, salmonellosis and tuberculosis. “Certified” milk, obtained from cows certified as healthy, is unpasteurized milk with a bacteria count below a specified standard, but it still can contain significant numbers of disease producing organisms. Different heat and treatments are given to raw milk in order to remove pathogenic organisms, to increase the shelf life, to help subsequent processing e.g. for warming before separation and homogenization or as an essential treatment before cheese making, yoghurt manufacture and production of evaporated and dried milk products (Singh, 1993).

Pasteurization, sterilization (in bottle) and UHT (ultra-high-temperature) treatment integrated with aseptic packing. Sterilization (in bottle) is the term applied to a heat treatment process which has a bactericidal effect greater than pasteurization. Although it does not result in sterility, it gives the processed milk a longer shelf life. As a result of the long holding time at this elevated temperature, the product has a cooked flavour and a pronounced brown colour. Unlike sterilization, pasteurization is not intended to kill all pathogenic micro-organisms in the food or liquid. Instead, pasteurization aims to reduce the number of viable pathogens so they are unlikely to cause disease.

Ultra-high temperature (UHT or ultra-heat treated) is also used for milk treatment. UHT processing holds the milk at a temperature of 138°C (250°F) for a fraction of a second. Milk simply labelled "pasteurization" is usually treated with the HTST method, whereas milk labelled "ultra-pasteurization" or simply "UHT" has been treated with the UHT method (Bylund, 1995). Heating of milk accounts 2 main problems, age gelation and off flavour development, which limits shelf life of milk. UHT treatment of milk leads to a much larger production of small sized casein micelles compared to raw or pasteurized milk (Singh, 1993). Biochemical processes involve are heat resistance and reactivation of natural and bacterial proteases and survival of bacterial spores.

Proteolysis of UHT milk during storage at room temperature is a major factor limiting the shelf life through changes in its flavour and texture (Datta et al., 2002). The changes ultimately reduce the quality and limit the shelf life of UHT milk via development of off flavours, fat separation and sedimentation, which principally falls into 2 categories, liberation of volatile fatty acids such as butyric acid and oxidation of free or unsaturated fatty acids (Datta et al., 2002). Above 135°C the protein deposited on the fat globule membrane form a network which makes the membrane denser and less permeable.

There is an increase in acidity and viscosity with a decrease in pH with the storage time increased both in UHT. Clare et al. (2005) determined that sweet aromatic flavour and sweet taste of UHT milk decreases during storage.

The microorganisms, which cause spoilage in milk, which is intended to be sterile (UHT treatment), are either resistant types that have survived the heat treatment, or organisms that have contaminated the product after the sterilization process. Contamination may either be by heat labile organism or heat resistant forms such as spores. Contaminating spores are, however, likely to be less heat resistant than those, which might survive the heat treatment.

The problem of post treatment contamination of in container sterilized product is well known. The contamination can either through poor seal or through pinhole in the container. Post treatment contaminants in UHT milk may be either spores, which would not be expected to be heat resistant enough to survive the heat treatment or non heat resistant vegetative organisms. Organisms of first type will probably have entered from ineffectively sterilized plant down stream from the heat treatment stage of the process, which includes spores of *Bacillus cereus* and *Bacillus licheniformis*. Organisms of second type will probably have entered through poorly sealed container after aseptic filling.

The types of spores, which have been investigated as of particular relevance in the UHT, are those of *Bacillus stearothermophilus*, *Bacillus subtilis* and *Clostridium botulinum* has been studied. The high spore counts can occur at the dairy farm and that feed and milking equipment can act as reservoirs or entry points for potentially highly heat resistant spores into raw milk.

Lowering this spore load by good hygienic measures could probably further reduce the contamination level of raw milk, in this way minimizing the aerobic spore forming bacteria that could lead to spoilage of milk and dairy products (Westhoff and Dougerty, 1981). These problems had been reported internationally since long, hence the project was planned to observe the physicochemical. In this study the de-clared shelf life of different UHT milk available in market is studied.

Materials and Methods

The samples were taken in sterilized syringes for microbiological analysis and in clean stainless steel containers of 1 liter for chemical and sensory analysis. The samples were analyzed at interval of 1, 2, 3, 4, 5, 6, 7, 8, 9, 10, 11 and 12 weeks. During this period, samples were stored at room temperature (25°C) to provide them similar conditions, as they are stored in market. For microbiological analysis the samples were examined for aerobic plate counts (APC), *E. coli* counts, *B. cereus, B. subtilis* counts and for spore formers counts. The parameters examined for the chemical analysis were sedimentation, pH, and acidity as lactic acid %, fat % before and after shaking the milk, SNF % before and after shaking and protein % before and after shaking. For sensory evaluation colour, aroma and taste were examined.

Physicochemical Analysis of Milk

To assess the physical and chemical changes in processed milk samples following tests were carried out.

Sedimentation Test

Sedimentation test was performed by following modified method as described by Ramsey and Swartzel (1984). According to this method, milk was drain from the cartons leaving the bottom 4 cm.

The cartons were inverted for approximately 10 min, up righted and placed in the exhaust hood to dry. The cartons were allowed to dry for 48 h after the bottom flaps or wings of cartons had been opened to facilitate the drying of any sediment entrapped there. The cartons were weighted and then washed thoroughly to remove any sediment or residue adhering to the container. The washed cartons were again dried and weighted.

Solids Non Fats (SNF) %

Solids non fats (SNF) % was determined by lactometeric method as described by Ramsey and Swartzel (1984).

Total Titratable Acidity

Total titratable acidity determined according to the method of AOAC, (2005).

pH

The pH value of milk was determined by using a digital pH meter (AOAC, 2005). Prior to use, the pH meter was standardized with standard buffer solution of pH 4 and 7.

Fat

Milk fat % was determined by Gerber (1997) method as described by FAO (1997) by using the butyrometer.

Protein

The protein was estimated by formal titration method (Davide, 1977).

Microbial Analysis

Microbial analysis was performed according to standard methods (AOAC, 2005).

Total Viable Counts

The plate count agar media (Bridson, 1995) was used for the total viable count in UHT milk samples (AOAC, 2005). Plates were incubated for 24 h at 37°C.

Determination of Coliforms

Coliform counts were determined by pour plate method on violet red bile agar, prepared according to the manufacturer instructions. All plates were incubated at 37°C for 24 h.

Determination of Bacillus Species

B. cereus selective agar base (Bridson, 1995) is used for isolation and enumeration of *B. cereus* and *B. subtilis*. All plates were incubated at 37°C for 24 h.

Determination of Spore Formers

Plate count agar media (Bridson, 1995) is used for the enumcration of spore formers. Sterile medium was poured into sterile petri plates and allowed the medium to solidify. Sample is heated at 80°C for 10 min using water bath. These plates were inoculated with 1 ml sample by using sterile pipette. After inoculation, the sample was well mixed in the petri plates by to and fro motion. All plates were incubated in an inverted position for 72 h at 55°C.

E. coli Counts

For *E. coli* count MacConkey's agar (Bridson, 1995) was used. Sample from lactose positive tubes in case of coliform counts were applied directly on the MacConkey's agar (Bridson, 1995) plates and incubated at 37°C for 24 h.

Sensory Analysis

The stored milk samples were evaluated sensorial for colour and flavor by scoring method as described by Larmond (1977).

Results and Discussion

The changes that have taken place during storage depend on temperature of storage, extent of exposure of the milk to light and availability of oxygen. The dairy company of Pakistan shows shelf life of 12 weeks on the labels of milk packs, during this mentioned period. Milk must be in best condition for consumption.

For storage time than a week or 2, these effects may be greater than those of the heat treatment. Changes in colour, flavour and texture are readily detected by the consumer and may reduce the acceptability of the products. Other changes cannot be recognized by the consumer and are not necessarily correlated with organoleptic, recognizable changes, but are of potential nutritional importance. The quality of sediment depends on the raw milk and on the type and severity of the heat treatments. For any 1 type of process, the amount of sediments increases in the severity of the heat treatment.

The amount of sediment decreases with homogenization pressure (Robinson, 1994). Results obtained from sedimentation test in UHT milk during storage period of 3 months (12 weeks) shows that there is an effect of heat processing and subsequent storage on sedimentation in all 4 samples of UHT milk. The changes started in week 2 of shelf life for samples I and III and sample II showed formation of sediments after week 6. Sample III reaches up to 7.10/250 gml-1, which is a considerable changes and sample II showed formation of sediments after week 5. The alcohol test can be used to detect raw milk that is likely to give high level of the normal type of sediments and there are indications that it may be useful in predicting the abnormal type (Sweetsur and White, 1975).

Processing operations influences acid base equilibrium in milk. UHT treatment results in a pH decrease, due to conversion of lactose into different organic acids. In milk, casein micelles are stable at natural pH, that is, 6.7. Lowering the pH facilitates aggregations of casein micelles and forms a gel. Results regarding effect of storage on pH of UHT processed milk during storage period of 90 days show that there is storage effect on pH level. Maximum pH value (6.81and 6.85 in samples 1 and 2, respectively and 6.75 in samples III and IV) was recorded in 1st week while minimum pH values obtained in 12th week of shelf life (6.20, 6.65, 6.17 and 6.55 in sample 1, 2, 3 and 4 of UHT milk respectively). Vankatachalm and McMahon (1991) verified drop in pH and they associated it with browning reactions. Andrews et al. (1977) confirmed similar effects and concluded that the level and extent of pH decrease was related to age gelation. When milk is heated at a temperature above 100oC and subsequent stored, lactose is degraded to acids. Formic acid is the principal acid produced due to which titratable acidity of milk rises.

Increase in free fatty acids is also responsible for increasing the total titratable acidity of milk (Swartzel, 1983). Results obtained by the analysis for total titratable acidity. The acidity value was 0.11% while during storage of UHT milk minimum acidity was recorded in 1st week and maximum value (0.18%) at 90 days life in sample 1 while 0.15 in case of sample 1 and 0.13 in samples 2 and 4. The proteins of milk are the constituents most affected by heating and subsequent storage of milk. The principal changes in UHT milk during storage may be due to enzymes. Many proteins in milk are very heat labile e.g. whey proteins, vitamin binding protein, antimicrobial proteins etc. These proteins coagulate after heating hence the texture of milk is deteriorated during storage (Fox and McSweeny, 1998).

Casein polymerization is greater at high storage temperature, but occurs significantly even under refrigeration: 50% of the protein may be in the polymer form after 6 months at 37°C, and 21% after 6 months at 4°C (Andrews, 1977). The results regarding protein % of stored UHT milk describes that there is effect of storage on protein contents of UHT processed milk. That is, in week 1 protein contents were 3.30%, 3.70% for sample 1 and 2 while in week 12 of storage were 2.35 and 3.48 respectively. In case of samples III and IV, protein contents were 3.40 in week 1 while it changes to 1.15 and 2.59, respectively, in week 12. There is no change in protein % in all samples after shaking of UHT milk. Chen et al. (2005) showed almost a 90% loss and denaturation of S-lactoglobulin (LG) of the UHT processed and dry milks by using polyacrylamide gel electrophoresis. Of the principle constituent, the fats are probably least affected by UHT treatment.

It is concluded from the whole study that there is an increase in sedimentation value, fat separation, titratable acidity during storage, while decrease was found in pH and protein % during storage of 12 weeks. The increase in sedimentation shows excessive protein denaturation during processing and subsequent storage. In UHT processed milks the fat separation was observed during storage.

This high % of fat separation is attributed with less homogenizing efficiency during processing. On microbiological examination, not any colony found on TPC plates, *coliform* agar plates, *E. coli* plate, *B. cereus, B. subtilus*, and spore formers plates, in all the 4 samples of UHT milk during storage of 12 weeks. Sensory characteristics showed a significant decrease in scores during storage. These all are factors that limit the shelf life of UHT milk.

The shelf life of milk mainly depend on the quality of raw milk and better quality of milk can be achieved in Pakistan, when the manufacturers have better milk collection system. The manufacturer of sample II has its own sophisticated type of milk collection system said to be VMCs (village milk collection centers). At these centers, milk is collected at small scale and in short time it is transported at low temperature to the processing plant, avoiding contamination, due to this practice the microbial as well as other contamination can be controlled in better way before heat treatment or processing. While manufacturers of other dairy industries of Pakistan get milk from contractors and ice added milk is mostly supplied to these industries which disturb the mineral balance and natural emulsion and give higher water activity which leads to physicochemical, microbiological as well as sensory changes during shelf life of milk.

Microbiological and Physicochemical Properties of Raw Milk Used for Processing Pasteurized Milk in Blue Nile Dairy Company (Sudan)

Milk is a highly nutritious food, ideal for microbial growth and the fresh milk easily deteriorates to become unsuitable for processing and human consumption (FAO 2001). High bacterial counts are indicator of poor production hygiene or ineffective pasteurization of milk (Harding 1999).

Milk and milk products derived from dairy cows milk can harbour a variety of microorganisms and can be important sources of foodborne pathogens. The presence of food-borne pathogens in milk is due to direct contact with contaminated sources in the dairy farm environment and to excretion from the udder of an infected animal (El Zubeir *et al.* 2006).

The hygienic quality problems of milk may arise from raw milk of diseased animals (Murphy and Boor 2000). Kang *et al.* (2005) reported that the presence of antimicrobial substances in raw milk could have serious toxicological and technical consequences. Raw milk may contain over 2,000,000 cfu/ml before processing of liquid milk or cheese making (Kameni *et al.* 2002). The raw milk distributed for consumption in Sudan does not find the real quality control measures needed to be of good quality food (Mohamed and El Zubeir 2007). However, some new private dairy plants started the processing of fluid milk and some dairy products. These are faced with many problems of which the quality control measures constitute an important concern. Hence, the present study was designed to assess the chemical, physical and microbial properties of raw milk supplied to the Blue Nile Dairy Company plant (CAPO) and to compare it with the produced pasteurized milk.

Materials and Methods

Source of Milk: This study was carried out during June to September 2005, and the raw milk for this study was collected from two dairies, namely Blue Nile Dairy Company and Kordi Farms. Blue Nile Dairy Company plant deals with both suppliers as a source of raw milk for processing pasteurized milk. Raw milk from both dairy farms is usually mixed before processing. Milk fat was standardized to 3-3.2%, and the milk was pasteurized at 72- 76< C for 15 second using a high temperature short time (HTST) plate heat exchanger (Wincantor Pasteurizer, Wincanton Engineering Ltd, South Street Sherborne Dorset, U.K.). The pasteurized milk was packed into Tetra Pack container (Tetra Pack Technical Services AB, Ruben Rausing gata, 5-221 86 Lund, Sweden).

Raw milk samples (36 samples) and pasteurized milk after processing and before packaging (12 samples) were examined for total bacterial counts (TBC), coliform, psychrotrophic (PC) and thermoduric bacterial counts. Physiochemical properties (fat, protein, lactose, ash, solid not fat, density, acidity, pH and freezing point), antibiotics and phosphatase test were also done.

Chemical Analysis of Milk Samples: The milk constituents (fat, protein, lactose, ash and solid not fat) and physical characteristics (density and freezing point), of the milk samples were determined by milk analyzer Lactoscan 90 (Aple Industries services–La Roche Sur Foron, France), according to manufacturer's instructions. Milk samples were mixed gently 4-5 times to avoid any air enclosure in the milk. Then 25 ml samples were taken in the sample-tube and put in the sample- holder one at a time with the analyzer in the recess position.

Then when the starting button activated, the analyzer sucks the milk, makes the measurements, returns the milk in the sample-tube and the digital indicator (IED display) shows the specified results.

Antibiotic residues were determined using Delvotest® SP- ampule Kit (202– Delvotest SP 100, test box, DSM Food Specialties, the Netherlands). The method was carried out according to the manufacturer's instructions. Phosphatase test was done using Lactognost tables and powders obtained from Heyl, Chem. Pharm-Fabrik, 14167 Berlin. The procedure for phosphatase test was done according to the manufacturer's instructions. The acidity of the samples was determined according to AOAC (1990). The temperature and the pH of the samples were determined using pH– meter (Wagtech, HI 8314 membrane pH Meter, U.K.).

Microbiological Analysis: The samples were examined for TBC, coliforms, thermoduric and psychrotrophic counts according to Houghtby *et al.* (1992); Christen *et al.* (1992); Ballou *et al.* (1995); Ravanis and Lewis (1995), respectively. Plate count agar No. 298 (Biomark Laboratories) was used for enumeration of TBC, thermoduric bacteria and psychrotrophic counts, while violet red bile agar No. 779 (Biomark Laboratories) was used to determine coliform counts. The media were prepared according to manufacturer's instructions. Plates for enumeration of TBC, thermoduric bacteria and coliforms were incubated at 32° C for 48 hours, 37° C for 48 hours and 37° C for 24 hours, respectively. Plates for enumeration of psychrotrophic counts were incubated at 7° C for ten days. Developed colonies were counted using manual colony counter. The plates counting 25-250 colonies were selected as described by Houghtby *et al.* (1992). The number reciprocal of the dilution factor was recorded as colony forming unit per ml (cfu/ml). The milk ring test (MRT) for brucellosis was carried out according to Harrigan and McCance (1976).

Statistical Analysis: Data were analyzed by SPSS programme (Statistical Package for Social Science, version 10.00). This test combines ANOVA with comparison of differences between means of the treatments at the significance level of $P < 0.05$.

Results and Discussion

The means of fat, protein, lactose, ash and solids not fat content were 4.14%, 3.48 %, 4.33%, 0.778% and 8.58% in raw milk samples mixture. The density, freezing point, titratable acidty and pH revealed 1.031, -0.520, 0.145 and 7.02. The analysis of variance showed highly significant variations ($P < 0.01$) due to the source of raw milk samples, except for fat. The composition of raw milks in the present study was compared favourably with the composition of milk in northern Europe, which contained fat of 4.3%, total protein of 3.4%, lactose of 4.65%, ash of 0.73%, TS of 13.3% and SNF of 9.0% (Invensys APV 2002). This result also agrees with that reported by El Zubeir *et al.* (2005) for raw milk. The present study revealed lower mean values for lactose (%) than that reported by El Zubeir *et al.* (2005). The lower lactose may be due to the effect of psychotrophic bacteria (Ballou *et al.* 1995). The results of physicochemical analysis of mixed raw milk used for producing fluid milk and the pasteurized milk. These results were higher compared to that reported by Elmagli and El Zubeir (2006a).

These differences in milk composition may be due to initial raw milk used and the procedure of processing. However the results of freezing point agreed with those reported by Tetra Pak Processing Systems (2003) for freezing points of raw and pasteurized milk -0.520± 0.001 and -0.447± 0.000, respectively obtained during the present study. This study also agreed with that reported by Elmagli and El Zubeir (2006a) who found the freezing point was -0.4734± 0.05032 C. The obtained data for acidity o of raw and pasteurized milk of 0.145% and 0.143%, respectively, which are in line with that reported by Harding (1999), while the mean value was lower than that reported by Mohamed and El Zubeir (2007). The microbiological quality of the raw milk used for processing pasteurized milk showed that the initial quality was good for TBC (log 4.800 cfu\ml), coliform counts (log 4.157 cfu/ml), thermoduric bacterial counts (log 2.994) and psychotrophic bacterial counts (log 810 cfu/ml). The analysis of variance showed highly significant differences (P< 0.01) due to the source of raw milk samples for TBC and coliform counts. This result was lower than that reported by PMO (2001) for the average standard plate counts for can and bulk milk (700.000 bacteria /ml and 100.0 bacteria /ml, respectively). Moreover, the microbial standards for grade A raw milk is 100.0 bacteria/ml (PMO, 2001). The lower counts of bacteria may be due to good cleaning system and good handling from farms to the plant as was stated before by Chye *et al.* (2004). Lower TBC value was obtained for pasteurized milk than that reported by Elmagli and El Zubeir (2006b), who reported a range of 6.5×105 to 6.5×1014, but was similar to that of Reena *et al.* (2003). In addition, PMO (2001) reported that the bacteria standards for grade A pasteurized milk should be less than 20,000 bacteria /ml. Coliform bacteria counts of pasteurized milk showed lower numbers than these reported by Elmagli and El Zubeir (2006b). The lower coliform counts might be due to hygienic quality of raw milk, proper pasteurization process, good packaging and good storage conditions. This agreed with PMO (2001) who reported that the total bacterial standards for grade A pasteurized milk should be < 10 coliform/ ml. In addition, coliform counts obtained are in line with Sudanese Standards (SSMO, 2005) which stated that the maximum coliform counts should not to exceed 102 cfu/ml.

Thermoduric bacterial counts (log 0.621cfu/ml) was lower than that reported by Mohamed and El Zubeir (2007). However, the present findings agreed with that reported by Invensys APV (2002) who reported an aerobic spores-forming bacteria of <400. The mean value of psychrotrophic bacteria for pasteurized milk was log 0.360 cfu/ml, which was lower counts compared with that reported by Elmagli and El Zubeir (2006b), who found the psychotrophic bacterial counts were <6.5×10 for pasteurized milk. All the samples during storage showed the absence of the phosphatase test. This result might be due to proper pasteurization. However, Elmagli and El Zubeir (2006a) demonstrated that 10 % of the pasteurized milk samples were positive to the phosphatase test. No brucella antibodies were detected in pasteurized milk, this might be due to proper pasteurization, and is in accord with that reported by OIE (2005). Moreover this result is better result than that reported by Alves *et al.* (2001). The presence of positive antibodies for brucella in the raw milk samples might suggest infection and/or vaccination, as those herds followed regular vaccination programmes. Similarly negative results of antibiotic residues test were obtained, this may be due to proper follow up of antibiotic withdrawal periods which indicated the good quality of raw milk used. These results are in agreement with Van Schaik *et al.* (2002) and Yamaki *et al.* (2004). It is concluded that the values of chemical contents are within standards limits except for lactose, whose value was lower than the reported by the dairy plant. Low TBC for pasteurized milk was obtained, and the results of this study clearly illustrate that pasteurization plays an important role in the survival and destruction of different bacterial contaminants.

Hazard Analysis Critical Control Point—HACCP

Hazard Analysis Critical Control Point or HACCP is a systematic preventive approach to food safety and pharmaceutical safety that addresses physical, chemical, and biological hazards as a means of prevention rather than finished product inspection. HACCP is used in the food industry to identify potential food safety hazards, so that key actions can be taken to reduce or eliminate the risk of the hazards being realized. The system is used at all stages of food production and preparation processes including packaging, distribution, etc. The Food and Drug Administration (FDA) and the United States Department of

Agriculture (USDA) say that their mandatory HACCP programs for juice and meat are an effective approach to food safety and protecting public health. Meat HACCP systems are regulated by the USDA, while seafood and juice are regulated by the FDA. The use of HACCP is currently voluntary in other food industries.

A forerunner to HACCP was developed in the form of production process monitoring during World War II because traditional "end of the pipe" testing was not an efficient way to ferret out artillery shells that would not explode. HACCP itself was conceived in the 1960s when the US National Aeronautics and Space Administration (NASA) asked Pillsbury to design and manufacture the first foods for space flights. Since then, HACCP has been recognized internationally as a logical tool for adapting traditional inspection methods to a modern, science-based, food safety system. Based on risk-assessment, HACCP plans allow both industry and government to allocate their resources efficiently in establishing and auditing safe food production practices. In 1994, the organization of *International HACCP Alliance* was established initially for the US meat and poultry industries to assist them with implementing HACCP and now its membership has been spread over other professional/industrial areas.

Hence, HACCP has been increasingly applied to industries other than food, such as cosmetics and pharmaceuticals. This method, which in effect seeks to plan out unsafe practices, differs from traditional "produce and test" quality control methods which are less successful and inappropriate for highly perishable foods. In the US, HACCP compliance is regulated by 21 CFR part 120 and 123. Similarly, FAO/WHO published a guideline for all governments to handle the issue in small and less developed food businesses.

History

On 4 October 1957, the Soviet Union launched Sputnik, the world's first satellite. American president Dwight D. Eisenhower responded by committing the United States to the space race. Eisenhower signed the National Aeronautics and Space Act on 29 July 1958 that created the National Aeronautics and Space Administration (NASA) to put an American satellite in orbit and to get a person in space.

Food played a critical part in the manned space program. The initial group involved in this were Herbert Hollander, Mary Klicka, and Hamed El-Bisi of the United States Army Laboratories in Natick, Massachusetts and Dr. Paul A. Lachance of the Manned Spaceflight Centre (Johnson Space Centre since February 1973) in Houston, Texas. Pillsbury joined the program as a contractor in 1959 with Howard E. Baumann representing the company as its lead scientist.

The main goal was to produce food that would not crumble under zero gravity, but also be safe to eat. Lachance imposed strict microbial requirements, including pathogen limits (including *E. coli*, *Salmonella*, and *Clostridium botulinum*) on all foods destined for space travel.

All personnel involved realized that traditional quality control methods would be inadequate because there would be so much product testing involved for actual product to be used. NASA own requirements for Critical Control Points (CCP) in engineering management would be used as a guide for food safety. CCP derived from Failure mode and effects analysis (FMEA) from NASA via the munitions industry to test weapon and engineering system reliability. Using that information, NASA and Pillsbury required contractors to identify "critical failure areas" and eliminate them from the system, a first in the food industry then. Baumann, a microbiologist by training, was so pleased with Pillsbury's experience in the space program that he advocated for his company to adopt what would become HACCP at Pillsbury.

Soon thereafter, Pillsbury was confronted with a food safety issue of its own when glass was found contaminated in farina, a cereal commonly used in infant food. Baumann's leadership promoted HACCP in Pillsbury for producing commercial foods, and applied to its own food production.

This led to a panel discussion at the 1971 National Conference on Food Protection that included examing CCPs and Good Manufacturing Practices in producing safe foods. Several botulism cases were attributed to under-processed low-acid canned foods in 1970-71. The United States Food and Drug Administration (FDA) asked Pillsbury to organize and conduct a training program on the inspection of canned foods for FDA inspectors. This 21 day program was first held in September 1972 with 11 days of classroom lecture and 10 days of canning plant evaluations. Canned food regulations (21 CFR 108, 21 CFR 110, 21 CFR 113, and 21 CFR 114) were first published in 1973. Pillsbury's training program to the FDA in 1972, titled "Food Safety through the Hazard Analysis and Critical Control Point System", was the first time that HACCP was used.

HACCP was initially set on three principles, now shown as principles one, two, and four in the section below. Pillsbury quickly adopted two more principles, numbers three and five, to its own company in 1975. It was further supported by the National Academy of Sciences (NAS) that governmental inspections by the FDA go from reviewing plant records to compliance with its HACCP system. A second proposal by the NAS led to the development of the National Advisory

Committee on Microbiological Criteria for Foods (NACMCF) in 1987. NACMCF was initially responsible for defining HACCP's systems and guidelines for its application and were coordinated with the Codex Committee for Food Hygiene, that led to reports starting in 1992 and further harmonization in 1997. By 1997, the seven HACCP principles listed below became the standard. A year earlier, the American Society for Quality offered their first certifications for HACCP Auditors. (First known as Certified Quality Auditor-HACCP, they were changed to Certified HACCP Auditor (CHA) in 2004.

HACCP expanded in all realms of the food industry, going into meat, poultry, seafood, dairy, and has spread now from the farm to the fork.

The HACCP Seven Principles

Principle 1: Conduct a hazard analysis. - Plans determine the food safety hazards and identify the preventive measures the plan can apply to control these hazards. A food safety hazard is any biological, chemical, or physical property that may cause a food to be unsafe for human consumption.

Principle 2: Identify critical control points. - A Critical Control Point (CCP) is a point, step, or procedure in a food manufacturing process at which control can be applied and, as a result, a food safety hazard can be prevented, eliminated, or reduced to an acceptable level.

Principle 3: Establish critical limits for each critical control point. - A critical limit is the maximum or minimum value to which a physical, biological, or chemical hazard must be controlled at a critical control point to prevent, eliminate, or reduce to an acceptable level.

Principle 4: Establish critical control point monitoring requirements. - Monitoring activities are necessary to ensure that the process is under control at each critical control point. In the United States, the FSIS is requiring that each monitoring procedure and its frequency be listed in the HACCP plan.

Principle 5: Establish corrective actions. - These are actions to be taken when monitoring indicates a deviation from an established critical limit. The final rule requires a plant's HACCP plan to identify the corrective actions to be taken if a critical limit is not met. Corrective actions are intended to ensure that no product injurious to health or otherwise adulterated as a result of the deviation enters commerce.

Principle 6: Establish record keeping procedures. - The HACCP regulation requires that all plants maintain certain documents, including its hazard analysis and written HACCP plan, and records documenting the monitoring of critical control points, critical limits, verification activities, and the handling of processing deviations.

Principle 7: Establish procedures for ensuring the HACCP system is working as intended. - Validation ensures that the plants do what they were designed to do; that is, they are successful in ensuring the production of a safe product. Plants will be required to validate their own HACCP plans. FSIS will not approve HACCP plans in advance, but will review them for conformance with the final rule.

Verification ensures the HACCP plan is adequate, that is, working as intended. Verification procedures may include such activities as review of HACCP plans, CCP records, critical limits and microbial sampling and analysis. FSIS is requiring that the HACCP plan include verification tasks to be performed by plant personnel. Verification tasks would also be performed by FSIS inspectors. Both FSIS and industry will undertake microbial testing as one of several verification activities. Verification also includes 'validation' - the process of finding evidence for the accuracy of the HACCP system (e.g. scientific evidence for critical limitations).

Standards

The seven HACCP principles are included in the international standard ISO 22000 FSMS 2005. This standard is a complete food safety and quality management system incorporating the elements of prerequisite programmes (GMP & SSOP), HACCP and the quality management system, which together form an organization's Total Quality Management system.

HACCP Training

HACCP management system trainings are only offered by several commercial enthusiasts. However, ASQ does provide Trained HACCP Auditor (CHA) exam to individuals seeking the professional training. In the UK the Chartered Institute of Environmental Health (CIEH) offer a HACCP for Food Manufacturing qualification accredited by the QCA (Qualifications and Curriculum Authority).

HACCP Application

Applied Range

It can apply to several food categories; sea food, bulk milk production line, Bulk Cream and Butter Production Line, animal meat industry, Organic Chemical Contaminants in Food, Corn Curl Manufacturing Plant and etc.

USA

- Fish and fishery products
- Fresh-cut produces

- Juice and nectary products
- Food outlets
- Meat and poultry products
- School food and services.

HACCP Implementation

It involves monitoring, verifying and validating of the daily work that is compliant with regulatory requirements in all stages all the time. The differences among those three types of work are given by Saskatchewan Agriculture and Food.

HACCP Versus ISO 22000

ISO 22000 is the new standard bound to replace HACCP on issues related to food safety. Although several companies, especially the big ones, have either implemented or are on the point of implementing ISO 22000, there are many others which are rather timid and/or reluctant to implement it. The main reason behind that is the lack of information and the fear that the new standard is too demanding in terms of bureaucratic work, from abstract of case study.

Emulsion

An emulsion is a mixture of two or more immiscible (unblendable) liquids. Emulsions are part of a more general class of two-phase systems of matter called colloids. Although the terms colloid and emulsion are sometimes used interchangeably, emulsion tends to imply that both the dispersed and the continuous phase are liquid. In an emulsion, one liquid (the dispersed phase) is dispersed in the other (the continuous phase).

Examples of emulsions include vinaigrettes, the photo-sensitive side of photographic film, milk and cutting fluid for metal working.

Structure and Properties of Emulsions

It is still common belief that emulsions basically do not display any structure, i.e., the droplets (or in case of dispersions, particles) dispersed in the liquid matrix (the "dispersion medium") are assumed to be statistically distributed. Therefore, for emulsions (like for dispersions) usually percolation theory is assumed to appropriately describe their properties.

However, percolation theory can only be applied if the system it should describe is in or close to thermodynamic equilibrium. There are very few studies about the structure of emulsions (dispersions), although they are plentiful in type and in use all over the world in innumerable applications.

In the following, only such emulsions will be discussed with a dispersed phase diameter of less than 1 μm. To understand the formation and properties of such emulsions (including dispersions), it must be considered, that the dispersed phase exhibits a "surface," which is covered ("wet") by a different "surface" which hence are forming an interface (chemistry). Both surfaces have to be created (which requires a huge amount of energy), and the interfacial tension (difference of surface tension) is not compensating the energy input, if at all.

A review article in introduces into various attempts to describe dispersions / emulsions. Dispersion is a process by which (in the case of solids becoming dispersed in a liquid) agglomerated particles are separated from each other and a new interface, between an inner surface of the liquid dispersion medium and the surface of the particles to be dispersed, is generated. Dispersion is a much more complicated (and less well understood) process than most people believe.

The above cited review article also displays experimental evidence for that dispersions have a structure very much different from any kind of statistical distribution (which would be characteristic for a system in thermodynamic equilibrium, but in contrast very much showing structures similar to self-organisation which can be described by non-equilibrium thermodynamics. This is the reason why some liquid dispersions turn to become gels or even solid at a concentration of a dispersed phase above a certain critical concentration (which is dependant on particle size and interfacial tension).

Also the sudden appearance of conductivity in a system of a dispersed conductive phase in an insulating matrix has been explained. The above cited review article also introduces into some first complete non-equilibrium thermodynamics theory of dispersions.

Appearance and Properties

Emulsions are made up of a dispersed and a continuous phase; the boundary between these phases is called the interface. Emulsions tend to have a cloudy appearance, because the many phase interfaces scatter light that passes through the emulsion.

Emulsions are unstable and thus do not form spontaneously. The basic colour of emulsions is white. If the emulsion is dilute, the Tyndall effect will scatter the light and distort the colour to blue; if it is concentrated, the colour will be distorted towards yellow. This phenomenon is easily observable on comparing skimmed milk (with no or little fat) to cream (high concentration of milk fat). Microemulsions and nanoemulsions tend to appear clear due to the small size of the disperse phase.

Energy input through shaking, stirring, homogenizing, or spray processes are needed to initially form an emulsion. Over time, emulsions tend to revert to the stable state of the phases comprising the emulsion; an example of this is seen in the separation of the oil and vinegar components of Vinaigrette, an unstable emulsion that will quickly separate unless shaken continuously.

Whether an emulsion turns into a water-in-oil emulsion or an oil-in-water emulsion depends on the volume fraction of both phases and on the type of emulsifier. Generally, the Bancroft rule applies: emulsifiers and emulsifying particles tend to promote dispersion of the phase in which they do not dissolve very well; for example, proteins dissolve better in water than in oil and so tend to form oil-in-water emulsions (that is they promote the dispersion of oil droplets throughout a continuous phase of water).

Instability

There are three types of instability: flocculation, creaming, and coalescence. Flocculation describes the process by which the dispersed phase comes out of suspension in flakes. Coalescence is another form of instability, which describes when small droplets combine to form progressively larger ones. Emulsions can also undergo creaming, the migration of one of the substances to the top (or the bottom, depending on the relative densities of the two phases) of the emulsion under the influence of buoyancy or centripetal force when a centrifuge is used. Surface active substances (surfactants) can increase the kinetic stability of emulsions greatly so that, once formed, the emulsion does not change significantly over years of storage. A Non-Ionic surfactant solution can become self-contained under the force of its own surface tension, remaining in the shape of its previous container for some time after the container is removed. Superfluids flow with zero friction and can escape their containers; an ionic solution tends to retain its current shape.

"Emulsion stability refers to the ability of an emulsion to resist change in its properties over time." D.J. McClements.

Technique Monitoring Physical Stability

Multiple light scattering coupled with vertical scanning is the most widely used technique to monitor the dispersion state of a product, hence identifying and quantifying destabilisation phenomena. It works on concentrated emulsions without dilution. When light is sent through the sample, it is backscattered by the droplets. The backscattering intensity is directly proportional to the size and volume fraction of the dispersed phase. Therefore, local changes in concentration (Creaming) and global changes in size (flocculation, coalescence) are detected and monitored.

Accelerating Methods for Shelf Life Prediction

The kinetic process of destabilisation can be rather long (up to several months or even years for some products) and it is often required for the formulator to use further accelerating methods in order to reach reasonable development time for new product design. Thermal methods are the most commonly used and consists in increasing temperature to accelerate destabilisation (below critical temperatures of phase inversion or chemical degradation). Temperature affects not only the viscosity, but also interfacial tension in the case of non-ionic surfactants or more generally interactions forces inside the system. Storing a dispersion at high temperatures enables to simulate real life conditions for a product (e.g. tube of sunscreen cream in a car in the summer), but also to accelerate destabilisation processes up to 200 times.

Mechanical acceleration including vibration, centrifugation and agitation are sometimes used. They subject the product to different forces that pushes the droplets against one another, hence helping in the film drainage. However, some emulsions would never coalesce in normal gravity, while they do under artificial gravity. Moreover segregation of different populations of particles have been highlighted when using centrifugation and vibration.

Emulsifier

An emulsifier (also known as an emulgent) is a substance which stabilizes an emulsion by increasing its kinetic stability. One class of emulsifiers is known as surface active substances, or surfactants. Examples of food emulsifiers are egg yolk (where the main emulsifying agent is lecithin), honey, and mustard, where a variety of chemicals in the mucilage surrounding the seed hull act as emulsifiers; proteins and low-molecular weight emulsifiers are common as well. Soy lecithin is another emulsifier and thickener. In some cases, particles can stabilize emulsions as well through a mechanism called Pickering stabilization. Both mayonnaise and Hollandaise sauce are oil-in-water emulsions that are stabilized with egg yolk lecithin or other types of food additives such as Sodium stearoyl lactylate.

Detergents are another class of surfactant, and will physically interact with both oil and water, thus stabilizing the interface between oil or water droplets in suspension. This principle is exploited in soap to remove grease for the purpose of cleaning. A wide variety of emulsifiers are used in pharmacy to prepare emulsions such as creams and lotions. Common examples include emulsifying wax, cetearyl alcohol, polysorbate 20, and ceteareth 20. Sometimes the inner phase itself can act as an emulsifier, and the result is nanoemulsion - the inner state disperses into nano-size droplets within the outer phase. A well-known example of this phenomenon, the ouzo effect, happens when water is poured in a strong alcoholic anise-based beverage, such as ouzo, pastis, arak or raki. The anisolic compounds, which are soluble in ethanol, now form nano-sized droplets and emulgate within the water. The colour of such diluted drink is opaque and milky.

In Food

Oil-in-water emulsions are common in food. Notable examples include:

- Crema in espresso – coffee oil in water (brewed coffee), unstable
- Hollandaise sauce – similar to mayonnaise
- Mayonnaise – vegetable oil in lemon juice or vinegar, with egg yolk lecithin as emulsifier
- Vinaigrette – vegetable oil in vinegar; if prepared with only oil and vinegar (without an emulsifier), yields an unstable emulsion.

In Medicine

In pharmaceutics, hairstyling, personal hygiene and cosmetics, emulsions are frequently used. These are usually oil and water emulsions, but which is dispersed and which is continuous depends on the pharmaceutical formulation. These emulsions may be called creams, ointments, liniments (balms), pastes, films or liquids, depending mostly on their oil and water proportions and their route of administration. The first 5 are topical dosage forms, and may be used on the surface of the skin, transdermally, ophthalmically, rectally or vaginally. A very liquidy emulsion may also be used orally, or it may be injected using various routes (typically intravenously or intramuscularly). Popular medicated emulsions include calamine lotion, cod liver oil, Polysporin, cortisol cream, Canesten and Fleet.

Microemulsions are used to deliver vaccines and kill microbes. Typically, the emulsions used in these techniques are nanoemulsions of soybean oil, with particles that are 400-600 nm in diameter. The process is not chemical, as with other types of antimicrobial treatments, but mechanical. The smaller the droplet, the greater the surface tension and thus the greater the force to merge with other lipids.

The oil is emulsified using a high shear mixer with detergents to stabilize the emulsion, so when they encounter the lipids in the membrane or envelope of bacteria or viruses, they force the lipids to merge with themselves. On a mass scale, this effectively disintegrates the membrane and kills the pathogen.

This soybean oil emulsion does not harm normal human cells nor the cells of most other higher organisms. The exceptions are sperm cells and blood cells, which are vulnerable to nanoemulsions due to their membrane structures. For this reason, these nanoemulsions are not currently used intravenously. The most effective application of this type of nanoemulsion is for the disinfection of surfaces. Some types of nanoemulsions have been shown to effectively destroy HIV-1 and various tuberculosis pathogens, for example, on non-porous surfaces.

In Fire Fighting

Emulsifying agents are effective at extinguishing fires on small thin layer spills of flammable liquids (Class B fires). Extinguishment is achieved by encapsulating the fuel in a fuel-water emulsion thereby trapping the flammable vapors in the water phase. This emulsion is achieved by applying an aqueous surfactant solution to the fuel through a high pressure nozzle. Emulsifiers are not effective at extinguishing large Class B fuel in depth fires. This is because the amount of agent needed for extinguishment is a function of the volume of the fuel whereas agents such as aqueous film forming foam (AFFF) need only cover the surface of the fuel to achieve vapor mitigation.

Uses

Emulsions are mainly used in many major chemical industries. In the pharmaceutical industry they are used to make medicines with a more appealing flavour and to improve value by controlling the amount of active ingredients. The most widely-used emulsions are non-ionic because they have low toxicity, but cationic emulsions are also used in some products because of their antimicrobial properties. Emulsions are also used in making many hair and skin products, such as various types of oils and waxes.

Shelf-life Predicting Methods for Milk

New analysis allows prediction of shelf life for pasteurized bottled milk. At I&A Lab we think that the most important part of a plant is production, and the laboratory should provide help to production to solve and prevent the development of quality problems. Production faces different types of problems like:

1- Quality of the raw milk received
2- Equipment
3- CIP
4- Packing material
5- Personnel.

All of these can affect the quality of the final product and as a consequence the shelf life.

Our analysis is designed for use with pasteurized milk. To establish the shelf life of bottled milk it is necessary to be able to predict how the product is going to behave in the stores, at the correct temperature, and to know what is going to happen if the product is abused. The ideal is to have this prediction in a reasonable amount of time, the sooner after the milk is bottled being better.

There are different groups of microorganisms capable of growing and spoiling the milk, for our study we consider that the mesophilic (growing between 68° and 113°F), and psychrophilic (growing between 20° and 68°F) are the most important groups, so we focus our effort on the mesophilic's which can spoil the milk when it is abused, and the psychrophilic's which can spoil the milk even at temperatures below 45°F.

Most psychrophilic microorganisms are a result of post - pasteurization contamination, due to the fact that they usually die with pasteurization. It is known that milk is an excellent media for the growth of many microorganisms, it is also known that if the milk is kept under 45°F many microorganisms stop or decrease their growth.

The legal analysis for pasteurized bottled milk uses one milliliter of milk in a solid media for total aerobic counts and coliform counts. It is not a requirement to analyze psychrophilics, but the most widely used analysis takes 7 to 10 days to complete.

The solid media employed is very different from liquid milk, so much so that it takes 48 hrs to see colonies growing in the solid media plates, incubated at 113°F, while it takes only a few hours for the milk to spoil. This means that for most of the microorganisms it is easier to grow in a liquid than in a solid. The reason for the use of solid media is to be able to count the amount of colonies per milliliter of milk.

We have developed a new method using liquid milk in a volume 10 times higher than normal in special equipment using new software. This allows us to detect microorganisms in very low concentrations. We also employ two different medias for psychrophilic bacteria, along with media for pseudomonas, and mesophilic bacteria. This allows us to have a wide spectrum of detectable milk inhabiting microorganisms.

We created a database with our results and compared them against the results of the plated fresh samples, and the same samples preincubated at 68°F for 16 hrs. Finally, the milk was flavoured over a period of time until it was determined to be no longer acceptable for consumption.

After more than 8000 samples being analyzed we are able to detect in 20 hrs samples which will have 10 days or less of shelf life. If the milk is contaminated for any reason we will be able to give the production plant this information. We are able to report in 24 hrs if the milk has any problem which may compromise its shelf life.

We will issue a second report in 48 hrs stating the amount of psychrophilic, pseudomonas, coliform, and mesophilic bacteria in the milk sample, with a prediction, in number of days, with which the milk will be in good condition if stored at 45°F. Also, the number of days of shelf life if the milk is kept between 50° and 55°F will be supplied.

During our study we found a type of bacteria capable of growing at 68°F. We identified this as Bacillus megaterium (mesophilic). As its presence occurred frequently we studied it and determined that even though it is capable of growing at room temperatures it is not capable of growing below 45°F, however it grows rapidly if the milk is abused.

As the milk is a biological product, and there are millions of different microorganisms capable of growing and making changes in the milk, we will continue our study. In the future we will be able to provide more information to help production managers make better decisions regarding the shelf life of their products.

Determination of the End of Shelf Life for Milk

Using Weibull Hazard Method

Undesirable changes in dairy products may be instigated by microbial growth and metabolism or by chemical reactions. The determinants of shelf life of fresh dairy products are usually the spoilage bacteria that have the ability to grow at refrigerated temperatures.

This microbial growth induces changes in the taste and odour of milk such as sour, putrid, bitter, malty, fruity, rancid and unclean. In addition, psychrotrophs which are common contaminants in milk, synthesize enzymes, many of which survive the pasteurization heat treatment and during storage biochemically alter the milk and eventually cause spoilage.

Growth of psychrotrophic bacteria is predominantly responsible for influencing the keeping quality of milk and dairy products held below 7°C. Raw and pasteurized milk usually spoils when held at refrigeration temperatures because of the effects of psychrotrophic contaminates.

The populations of microorganisms needed to cause detectable changes in milk varies among genera and species within a genus, but levels at which flavour changes occur are similar at 6 and 20°C. Milk spoilage by psychrotrophs was reported in the range of populations of 1 x 102 to 1 x 109 per ml. It is therefore unclear whether pschrotrophs counts can be used as an index in the determination of milk quality or shelf life from a sensory standpoint.

As noted earlier, microbial spoilage leads to sensory deterioration of the milk. It may therefore be suggested that the microbial quality of the milk should correlate well to its sensory end of shelf life. The end of shelf life can be determined from sensory data by various graphical methods. The use of hazard rate for shelf life testing of food was introduced by Gacula (1975). Using this method, one can determine the end of shelf life according to the percent of customers a company is prepared to displease. The maximum likelihood graphical procedure, or Weibull Hazard method has been used for shelf life of luncheon meats, oat bran cereal, ice cream, cottage cheese, Bockwurst sausages and butter, and other food products.

The objective of this research was to determine whether or not a consumer determined end of sensory shelf life can be described by some microbial index regardless of the temperature conditions that the milk is stored at.

Materials and Methods

Milk

The milk used in this study was TLC¨ fat free milk with added Calcium. This milk is also fortified with nonfat milk proteins. The raw milk was held for no longer than 48 hours at 2°C before processing and then pasteurized (20 s, 80°C). Half gallon, paper board cartons of milk were picked up within 2 h after bottling, taken off the production line consecutively by a plant supervisor, in order to minimize variability between cartons. The cartons of milk were transported to the University of Minnesota on ice in a cooler. Immediately after arrival, the milk was sampled for microbial quality and tasted by three expert dairy panelists to ensure that the milk was of good quality.

Microbial Counts

Total aerobic bacteria as well as psychrotrophic bacteria were enumerated using 3M Petrifilm (3M Co., St. Paul, MN). Samples were diluted in 0.1% peptone, and 1 ml of sample was transfered onto the film in duplicate. The Petrifilm contained standard method nutrients and a cold water soluble gelling agent (8, 20). The bottom film is coated with nutrients and gelling agent, while the top film is coated with the gelling agent and 2,3, 5-triphenylterazoluim chloride (TTC). Colonies appear red and were counted following incubation at 37°C for 48 h for total aerobic or 10 d at 7°C for pyschrotrophs.

Microbial Growth

The growth of total aerobic bacteria and psychrotrophic bacteria in the milk was measured at five constant temperatures: 2, 5, 7, 12, and 15°C (± 1°C). A TempTale (Sensitech, Beverly, MA) temperature recorder, placed in the coolers along with the milk verified the temperature history. Samples were drawn at predetermined intervals according to the storage temperature conditions. The lag times were determined graphically, and the growth rate constant was calculated through linear regression of the exponential phase of the growth curves.

Sensory Testing

The sensory testing was carried out following the Weibull Hazard method, where the initial number of panelists was n0 = 3 and the constant with which the number of panelists was increased for each subsequent test was nc = 1. The interval between sensory testing was predetermined for each of the five different storage conditions.

Because the spoilage of milk at 14°C occurred at an accelerated rate, sensory samples were held overnight so that the sensory test could be carried out at a convenient time for the panelists. The panelists were prescreened and were required to meet the criterion that they consume at least one 8-ounce glass of milk a day. A pool of 33 panelists who met this requirement, 16 male and 17 female, ranging in age form 18 to 45 were available for sensory testing. The panelists were financially compensated according to the number of samples that they tested.

For each sensory test a sample of milk was taken from the milk cartons, poured into a glass flask and the flask was immediately placed into an ice bath to slow down any microbial growth that might have caused any further deterioration of the sensory quality of the milk. Approximately 10 ml of milk was poured into cups that were labelled with random three digit numbers for identification purposes. A tray of milk samples for each of the panelists was prepared approximately half an hour to an hour before sensory testing took place.

To ensure that the samples were all the same temperature when the panelists received their trays the trays were stored in conventional home refrigerators held at 4°C. The trays, consisted of samples of milk from the different storage temperatures. The trays were presented to the panelists in sensory booths where the sensory testing was held. Panelists were asked to taste the first sample and determine whether the milk was acceptable or unacceptable, where a response of acceptable implied that the panelist would be willing to drink an entire glass of the sample. Panelists were asked to wait two minutes between samples and to rinse their mouths with water in between.

The end of sensory shelf life was determined at 69.3 % cumulative hazard or a critical probability of 50%. The data was regressed using the least squares method up to 100% cumulative hazard.

Results and Discussion

Microbial Growth

The growth of the total aerobic bacteria and the psychrotrophic bacteria were obtained at 2, 5, 7, 12, and 14°C (± 1°C).

The total aerobic microbial population exhibited typical growth curves at all temperatures, however, at 2 and 5°C the milk reached the end of sensory shelf life before the end of the lag phase. The psychrotrophic bacteria, exhibit distinct lag and log growth phases at 5, 7, 12, and 14°C. At 2°C however, a rapid growth of psychrotrophs occurred immediately following the opening of the milk carton for sampling, thus, obtaining a growth curve for psychrotrophs at 2°C was not practical within this experimental setup. It is likely that the rapid growth following opening of the milk carton was due to the change in available oxygen. Once the carton is opened the amount of available oxygen for the microorganisms in the milk increases and facilitates their growth. Sinclair and Stokes (1963) support this explanation with the discovery that in general, due to an increase in the availability of oxygen higher counts are observed.

The growth of both aerobic and pschrotrophs populations at 5°C did not exhibit a distinct logarithmic phase. The absence of a logarithmic growth phase at 5°C can be attributed to the fact that the samples were taken from more than one carton throughout the experiment. Thus the variability in the population of microorganisms from carton to carton may result in difficulties in detecting a distinct lag and exponential phase.

The cartons were taken from the production line in consecutive order so that the cartons could be considered to be identical and so the sampling from the cartons during the growth curve study could be made randomly between the opened cartons. However, Maxcy and

Wallen (1983) found that heterogeneity between cartons was apparent even when samples were taken sequentially from a single production line.

Growth Parameters

It was not possible to determine the duration of the lag phase for psychrotrophic bacteria of the milk stored at 2°C because the psychrotrophic counts showed no distinctive pattern. Since the milk stored at 5° and 2°C reached the sensory end point during the lag phase of both the total aerobic bacteria and psychrotrophic bacteria, the exponential growth rates for the bacteria at these two storage temperatures, were not calculated. Data presented by Fu (1989) showed that the temperature dependence of the growth rate constants and the lag time for microbial growth in a model milk system fits the Arrhenius model:

Indigenous (Indian) Dairy Products

A variety of dairy projects are indigenous to India and an important part of Indian cuisine. The majority of these products can be broadly classified into curdled products, like chhena, or non-curdled products, like khoa.

Curdled Dairy Products

- Paneer is an unaged, acid-set, non-melting farmer cheese made by curdling heated milk with lemon juice or other non-rennet food acid, and then removing the whey and pressing the result into a dry unit.
- Chhena is like paneer, except some whey is left and the mixture is beaten thoroughly until it becomes soft, of smooth consistency, and malleable but firm.
- Sandesh is a confection made from chhena mixed with sugar then grilled lightly to caramelize, but removed from heat and molded into a ball or some shape.
- Rasgulla is a confection made from mixture of chhena and semolina rolled into a ball and boiled in syrup.

Non-curdled Dairy Products

- Khoa or Mawa is made by reducing milk in an open pan over heat.
- Peda is a confection made by mixing sugar with khoa and adding flavoring, such as cardamom.
- Barfi is a confection made by reducing milk and sugar until it solidifies and adding flavoring, such as pistachio.

- Gulab jamun is a confection made by mixing khoa and sugar, caramelizing it by frying, and soaking it in syrup containing rosewater.
- Kulfi is made from slowly freezing sweetened condensed milk. In comparison to ice cream, kulfi is not whipped or otherwise aerated.
- Ghee is type of clarified butter that is cooked long enough to caramelize the milk sugar and sterilize the liquid.

Fermented Dairy Products

- Mishti doi is *dahi* (Indian yogurt) mixed with sugar
- Shrikhand is strained yoghurt mixed with sugar, and often flavourings such as cardamom, saffron, or fruit.
- Wheyvit is an alcoholic beverage prepared by fermenting whey with yeast.

Other Dairy Products

- Kheer is made by boiling rice or broken wheat with milk and sugar, and sometimes flavoured with cardamom, raisins, saffron, pistachios, or almonds.
- Chhena Murki is made by frying cubes of chhena to burn the outside, then soaking them in syrup flavoured with cardamom.
- Pantooa is like gulab jamun, except with some chhena mixed with the usual ingredients.

Chapter 5

Fluid Milk Processing

Beverage Milks

The production of beverage milks combines the unit operations of clarification, separation (for the production of lower fat milks), pasteurization, and homogenization. The process is simple, as indicated in the flow chart. While the fat content of most raw milk is 4% or higher, the fat content in most beverage milks has been reduced to 3.4%. Lower fat alternatives, such as 2% fat, 1% fat, or skim milk (<0.1% fat) or also available in most markets. These products are either produced by partially skimming the whole milk, or by completely skimming it and then adding an appropriate amount of cream back to achieve the desired final fat content.

Vitamins may be added to both full fat and reduced fat milks. Vitamins A and D (the fat soluble ones) are often supplemented in the form of a water soluble emulsion to offset that quantity lost in the fat separation process.

Creams

During the separation of whole milk, two streams are produced: the fat-depleted stream, which produces the beverage milks as described above or skim milk for evaporation and possibly for subsequent drying, and the fat-rich stream, the cream. This usually comes off the separator with fat contents in the 35-45% range. Cream is used for further processing in the dairy industry for the production of ice cream or butter, or can be sold to other food processing industries. These industrial products normally have higher fat contents than creams for retail sale, normally in the range of 45-50% fat. A product known as "plastic" cream can be produced from certain types of milk separators. This product has a fat content approaching 80% fat, but it remains as an oil-in-water emulsion (the fat is still in the form of globules and the skim milk is the continuous phase of the emulsion), unlike butter which also has a fat content of 80% but which has been churned so that the fat occupies the continuous phase and the skim milk is dispersed throughout in the form of tiny droplets (a water-in-oil emulsion).

For retail cream products, the fat is normally standardized to 35% (heavy cream for whipping), 18% or 10% (cream for coffee or cereal). Higher fat creams have also been produced for retail sale, a product known as double cream has a fat content of 55% and is quite thick. Creams for packaging and sale in the retail market must be pasteurized to ensure freedom from pathogenic bacteria. Whipping cream is not normally homogenized, as the high fat content will lead to extensive fat globule aggregation and clustering, which leads to excessive viscosity and a loss of whipping ability. This phenomena has been used, however, to produce a spoonable cream product to be used as a dessert topping. Lower fat creams (10% or 18%) can be homogenized, usually at lower pressure than whole milk.

Recombined Milk

Beverage milks can also be prepared by recombining skim milk powder and butter with water. This is often done in countries where there is not enough milk production to meet the demand for beverage milk consumption. The concept is simple. Skim milk powder is dispersed in water and allowed to hydrate. Butter is then emulsified into this mixture by either blending melted butter into the liquid mixture while hot, or by dispersing solid butter into the liquid through a high shear blender device.

In some cases, a non-dairy fat source may also be used. The recombined milk product is then pasteurized, homogenized and packaged as in regular milk production. The final composition is similar to that of whole milk, approximately 9% milk solids-not-fat, and either 2% or 3.4% fat. The water source must be of excellent quality. The milk powder used for recombining must be of high quality and good flavour. Care must be taken to ensure adequate blending of the ingredients to prevent aggregation or lumping of the powder. Its dispersal in water is the key to success.

Chocolate Milk

An industry standard for the production of chocolate milk consists of:

- 93% milk
- 6.3% sugar

- 0.65% cocoa powder
- 0.05% carrageenan.

The final product is usually standardized to either 2% fat or 1% fat (meaning, 2.15% or 1.1% fat in the milk before addition of other ingredients). The sugar, cocoa powder and carrageenan are dry blended, and added to cold milk with vigourous agitation, and then pasteurized.

Concentrated and Dried Dairy Products

Fluid milk contains approximately 88% water. Concentrated milk products are obtained through partial water removal. Dried dairy products have even greater amounts of water removed to usually less than 4%. The benefits of both these processes include an increased shelf-life, convenience, product flexibility, decreased transportation costs, and storage.

The following products will be discussed here:

- Concentrated Dairy Products
- Evaporated Skim or Whole Milk.

After the raw milk is clarified and standardized, it is given a pre-heating treatment of 93-100° C for 10 to 25 min or 115-128° C for 1 to 6 min.. There are several benefits to this treatment:

- increases the concentrated milk stability during sterilization; decreases the chance of coagulation taking place during storage
- decreases the initial microbial load
- modifies the viscosity of the final product
- milk enters the evaporator already hot.

Milk is then concentrated at low temperatures by vacuum evaporation. This process is based on the physical law that the boiling point of a liquid is lowered when the liquid is exposed to a pressure below atmospheric pressure. In this case, the boiling point is lowered to approximately 40-45° C. This results in little to no cooked flavour. The milk is concentrated to 30-40% total solids. The evaporated milk is then homogenized to improve the milkfat emulsion stability. There are other benefits particular to this type of product:

- increased white colour
- increased viscosity
- decreased coagulation ability.

A second standardization is done at this time to ensure the proper salt balance is present. The ability of milk to withstand intensive heat treatment depends to a great degree on its salt balance.

The product at this point is quite perishable. The fat is easily oxidized and the microbial load, although decreased, is still a threat. The evaporated milk at this stage is often shipped by the tanker for use in other products.

In order to extend the shelf life, evaporated milk can be packaged in cans and then sterilized in an autoclave. Continuous flow sterilization followed by packaging under aseptic conditions is also done. While the sterilization process produces a light brown colouration, the product can be successfully stored for up to a year.

Sweetened Condensed Milk

Where evaporated milk uses sterilization to extend its shelf-life, sweetened condensed milk has an extended shelf-life due to the addition of sugar. Sucrose, in the form of crystals or solution, increases the *osmotic pressure* of the liquid. This in turn, prevents the growth of microorganisms.

The only real heat treatment (85-90° C for several seconds) this product recieves is after the raw milk has been clarified and standardized. The benefits of this treatment include totally destroying osmophilic and thermophilic microorganisms, inactivating lipases and proteases, decreases fat separation and inhibits oxidative changes. Unfortunately it also affects the final product viscosity and may promote the defect age gelation. The milk is evaporated in a manner similar to the evaporated milk. Although sugar may be added before evaporation, post evaporation addition is recommended to avoid undesirable viscosity changes during storage.

Enough sugar is added so that the final concentration of sugar is approximately 45%.

The sweetened evaporated milk is then cooled and lactose crystallization is induced. The milk is inoculated, or seeded, with powdered lactose crystals, then rapidly cooled while being agitated. The lactose can crystalize without the seeding but there is the danger of forming crystals that are too large. This would result in a texture defect similar in ice cream called sandiness, which affects the mouthfeel. By seeding, the number of crystals increases and the size of those crystals decreases.

The product is packaged in smaller containers, such as cans, for retail sales and bulk containers for industrial sales.

Condensed Buttermilk

Buttermilk is a by-product of the butter industry. It can be evaporated on its own or it can be blended with skimmilk and dried to produce skimmilk powder. This blended product may oxidise readily due to the higher fat content. Condensed buttermilk is perishable and, therefore, the supply must be fresh and it must be stored cool.

Condensed Whey

In the process of cheesemaking, there is alot of whey that needs to be disposed of. One of the ways of utilizing cheesewhey is to condense it. The whey contains fat, lactose, β-lactoglobulin, alpha-lactalbumin, and water. The fat is generally removed by centrifugation and churned as whey cream or used in ice cream. Evaporation is the first step in producing whey powder.

Dried Dairy Products

Milk Powder

Milk used in the production of milk powders is first clarified, standardized and then given a heat treatment. This heat treatment is usually more severe than that required for pasteurization. Besides destroying all the pathogenic and most of the spoilage microorganisms, it also inactivates the enzyme lipase which could cause lipolysis during storage.

The milk is then evaporated prior to drying for the following reasons:

- less occluded air and longer shelf life for the powder
- viscosity increase leads to larger powder particles
- less energy required to remove part of water by evaporation; more economical.

Homogenization may be applied to decrease the free fat content. Spray drying is the most used method for producing milk powders. After drying, the powder must be packaged in containers able to provide protection from moisture, air, light, etc. Whole milk powder can then be stored for long periods (up to about 6 months) of time at ambient temperatures.

Skim milk powder (SMP) processing is similar to that described above except for the following points:

1. contains less milkfat (0.05-0.10%)
2. heat treatment prior to evaporation can be more or less severe
3. homogenization not required
4. maximum shelf life extended to approximately 3 years.

Low-heat SMP is given a pasteurization heat treatment and is used in the production of cheese, baby foods etc. High-heat SMP requires a more intense heat treatment in addition to pasteurization. This product is used in the bakery industry, chocolate industry, and other foods where a high degree of protein denaturation is required.

Instant milk powder is produced by partially rehydrating the dried milk powder particles causing them to become sticky and agglomerate. The water is then removed by drying resulting in an increased amount of air incorporated between the powder particles.

Whey Powder

Whey is the by-product in the manufacturing of cheese and casein. Disposing of this whey has long been a problem. For environmental reasons it cannot be discharged into lakes and rivers; for economical reasons it is not desirable to simply dump it to waste treatment facilities. Converting whey into powder has led to a number products that it can be incorporated into.

It is most desirable, if and where possible, to use it for human food, as it contains a small but valuable protein component. It is also feasible to use it as animal feed. Between the pet food industry and animal feed mixers, hundred's of millions of pounds are sold every year. The feed industry may be the largest consumer of dried whey and whey products.

Whey powder is essentially produced by the same method as other milk powders. Reverse osmosis can be used to partially concentrate the whey prior to vacuum evaporation. Before the whey concentrate is spray dried, lactose crystallization is induced to decrease the hygroscopicity. This is accomplished by quick cooling in flash coolers after evaporation. Crystallization continues in agitated tanks for 4 to 24 h.

A fluidized bed may be used to produce large agglomerated particles with free-flowing, non-hygroscopic, no caking characteristics.

Whey Protein Concentrates

Both whey disposal problems and high-quality animal protein shortages have increased worldwide interest in whey protein concentrates. After clarification and pasteurization, the whey is cooled and held to stabilize the calcium phosphate complex, which later decreases membrane fouling.

The whey is commonly processed using ultrafiltration, although reverse osmosis, microfiltration, and demineralization methods can be used. During ultrafiltration, the low molecular weight compounds such as lactose, minerals, vitamins and nonprotein nitrogen are removed in the permeate while the proteins become concentrated in the retentate. After ultrafiltration, the retentate is pasteurized, may be evaporated, then dried. Drying, usually spray drying, is done at lower temperatures than for milk in order that large amounts of protein denaturation may be avoided.

Cheese

Traditionally, cheese was made as a way of preserving the nutrients of milk. In a simple definition, cheese is the fresh or ripened product obtained after coagulation and whey separation of milk, cream or partly skimmed milk, buttermilk or a mixture of these products. It is essentially the product of selective concentration of milk. Thousands of varieties of cheeses have evolved that are characteristic of various regions of the world.

Treatment of Milk for Cheesemaking

Like most dairy products, cheesemilk must first be clarified, separated and standardized. The milk may then be subjected to a sub-pasteurization treatment of 63-65° C for 15 to 16 sec. This thermization treatment results in a reduction of high initial bacteria counts before storage. It must be followed by proper pasteurization. While HTST pasteurization (72° C for 16 sec) is often used, an alternative heat treatment of 60° C for 16 sec may also be used. This less severe heat treatment is thought to result in a better final flavour cheese by preserving some of the natural flora. If used, the cheese must be stored for 60 days prior to sale, which is similar to the regulations for raw milk cheese.

Homogenization is not usually done for most cheesemilk. It disrupts the fat globules and increases the fat surface area where casein particles adsorb. This results in a soft, weak curd at renneting and increased hydrolytic rancidity.

Additives

The following may all be added to the cheese milk:

- Calcium choride
- nitrates
- colour
- hydrogen peroxide
- lipases.

Calcium choride is added to replace calcium redistributed during pasteurization. Milk coagulation by rennet during cheese making requires an optimum balance among ionic calcium and both soluble insoluble calcium phosphate salts. Because calcium phosphates have reverse solubility with respect to temperature, the heat treatment from pasteurization causes the equilibrium to shift towards insoluble forms and depletes both soluble calcium phosphates and ionic calcium. Near normal equilibrium is restored during 24-48 hours of cold storage, but cheese makers can't wait that long, so $CaCl_2$ is added to restore ionic calcium and improve rennetability. The calcium assists in coagulation and reduces the amount of rennet required.

Sodium or potassium nitrate is added to the milk to control the undesirable effects of *Clostridium tyrobutyricum* in cheeses such as Edam, Gouda, and Swiss.

Because milk colour varies from season to season, colour may added to standardize the colour of the cheese throughout the year. Annato, Beta-carotene, and paprika are used. The addition of hydrogen peroxide is sometimes used as an alternative treatment for full pateurization.

Lipases, normally present in raw milk, are inactivated during pasteurization. The addition of kid goat lipases are common to ensure proper flavour development through fat hydrolysis.

Inoculation and Milk Ripening

The basis of cheesemaking relies on the fermentation of lactose by lactic acid bacteria (LAB). LAB produce lactic acid which lowers the pH and in turn assists coagulation, promotes syneresis, helps prevent spoilage and pathogenic bacteria from growing, contributes to cheese texture, flavour and keeping quality. LAB also produce growth factors which encourages the growth of non-starter organisms, and provides lipases and proteases necessary for flavour development during curing. Further information on LAB and starter cultures can be found in the microbiology section.

After innoculation with the starter culture, the milk is held for 45 to 60 min at 25 to 30° C to ensure the bacteria are active, growing and have developed acidity. This stage is called ripening the milk and is done prior to renneting.

Milk Coagulation

Coagulation is essentially the formation of a gel by destabilizing the casein micelles causing them to aggregate and form a network which partially immobilizes the water and traps the fat globules in the newly formed matrix. This may be accomplished with:

- enzymes
- acid treatment
- heat-acid treatment.

Enzymes

Chymosin, or rennet, is most often used for enzyme coagulation.

Acid Treatment

Lowering the pH of the milk results in casein micelle destabilization or aggregation. Acid curd is more fragile than rennet curd due to the loss of calcium. Acid coagulation can be achieved naturally with the starter culture, or artificially with the addition of gluconodeltalactone. Acid coagulated fresh cheeses may include Cottage cheese, Quark, and Cream cheese.

Heat-Acid Treatment

Heat causes denaturation of the whey proteins. The denatured proteins then interact with the caseins. With the addition of acid, the caseins precipitate with the whey proteins. In rennet coagulation, only 76-78% of the protein is recovered, while in heat-acid coagulation, 90% of protein can be recovered. Examples of cheeses made by this method include Paneer, Ricotta and Queso Blanco.

Curd Treatment

After the milk has gel has been allowed to reach the desired firmness, it is carefully cut into small pieces with knife blades or wires. This shortens the distance and increases the available area for whey to be released. The curd pieces immediately begin to shrink and expel the greenish liquid called whey. This syneresis process is further driven by a cooking stage. The increase in temperature causes the protein matrix to shrink due to increased hydrophobic interactions, and also increases the rate of fermentation of lactose to lactic acid. The increased acidity also contributes to shrinkage of the curd particles. The final moisture content is dependant on the time and temperature of the cook stage. This is important to monitor carefully because the final moisture content of the curd determines the residual amount of fermentable lactose and thus the final pH of the cheese after curing.

When the curds have reached the desired moisture and acidity they are separated from the whey. The whey may be removed from the top or drained by gravity. The curd-whey mixture may also be placed in moulds for draining. Some cheese varieties, such as Colby, Gouda, and Brine Brick include a curd washing which increases the moisture content, reduces the lactose content and final acidity, decreases firmness, and increases openness of texture.

Curd handling from this point on is very specific for each cheese variety. Salting may be achieved through brine as with Gouda, surface salt as with Feta, or vat salt as with Cheddar. To achieve the characteritics of Cheddar, a cheddaring stage (curd manipulation), milling (cut into shreds), and pressing at high pressure are crucial.

Cheese Ripening

Except for fresh cheese, the curd is ripened, or matured, at various temperatures and times until the characteristic flavour, body and texture profile is achieved. During ripening, degradation of lactose, proteins and fat are carried out by ripening agents. The ripening agents in cheese are:

- bacteria and enzymes of the milk
- lactic culture
- rennet
- lipases
- added moulds or yeasts
- environmental contaminants.

Thus the microbiological content of the curd, the biochemical composition of the curd, as well as temperature and humidity affect the final product. This final stage varies from weeks to years according to the cheese variety.

Yogurt

Yogurt (also spelled yogourt or yoghurt) is a semi-solid fermented milk product that originated centuries ago and has evolved from many traditional Eastern European (e.g., Turkish and Bulgarian) products. The word is from the Turkish *Yogen*, meaning *thick*. It's popularity has grown and is now consumed in most parts of the world.

Ingredients

Although milk of various animals has been used for yogurt production in various parts of the world, most of the industrialized yogurt production uses cow's milk. Whole milk, partially skimmed milk, skim milk or cream may be used. In order to ensure the development of the yogurt culture the following criteria for the raw milk must be met:

- low bacteria count
- free from antibiotics, sanitizing chemicals, mastitis milk, colostrum, and rancid milk
- no contamination by bacteriophages.

Other yogurt ingredients may include some or all of the following:

Other Dairy Products: concentrated skim milk, nonfat dry milk, whey, lactose. These products are often used to increase the nonfat solids content.

Sweeteners: glucose or sucrose, high-intensity sweeteners (e.g. aspartame).

Stabilizers: gelatin, carboxymethyl cellulose, locust bean Guar, alginates, carrageenans, whey protein concentrate.

Flavours

Fruit Preparations: including natural and artificial flavouring, colour.

Starter Culture

The starter culture for most yogurt production in North America is a symbiotic blend of *Streptococcus salivarius* subsp. *thermophilus* (ST) and *Lactobacillus delbrueckii* subsp. *bulgaricus* (LB). Although they can grow independantly, the rate of acid production is much higher when used together than either of the two organisms grown individually. ST grows faster and produces both acid and carbon dioxide. The formate and carbon dioxide produced stimulates LB growth. On the other hand, the proteolytic activity of LB produces stimulatory peptides and amino acids for use by ST. These microorganisms are ultimately responsible for the formation of typical yogurt flavour and texture. The yogurt mixture coagulates during fermentation due to the drop in pH. The streptococci are responsible for the initial pH drop of the yogurt mix to approximately 5.0. The lactobacilli are responsible for a further decrease to pH 4.0. The following fermentation products contibute to flavour:

- lactic acid
- acetaldehyde
- acetic acid
- diacetyl.

Manufacturing Method

The milk is clarified and separated into cream and skim milk, then standardized to achieve the desired fat content. The various ingredients are then blended together in a mix tank equipped with a powder funnel and an agitation system. The mixture is then pasteurized using a continuous plate heat exchanger for 30 min at 85° C or 10 min at 95° C. These heat treatments, which are much more severe than fluid milk pasteurization, are necessary to achieve the following:

- produce a relatively sterile and conducive environment for the starter culture
- denature and coagulate whey proteins to enhance the viscosity and texture.

The mix is then homogenized using high pressures of 2000-2500 psi. Besides thoroughly mixing the stabilizers and other ingredients, homogenization also prevents creaming and wheying off during incubation and storage. Stability, consistency and body are enhanced by homogenization. Once the homogenized mix has cooled to an optimum growth temperature, the yogurt starter culture is added.

A ratio of 1:1, ST to LB, inoculation is added to the jacketed fermentation tank. A temperature of 43° C is maintained for 4-6 h under quiescent (no agitation) conditions. This temperature is a compromise between the optimums for the two micoorganisms (ST 39° C; LB 45° C). The titratable acidity is carefully monitored until the TA is 0.85 to 0.90%. At this time the jacket is replaced with cool water and agitation begins, both of which stop the fermentation. The coagulated product is cooled to 5-22° C, depending on the product. Fruit and flavour may be incorporated at this time, then packaged. The product is now cooled and stored at refrigeration temperatures (5° C) to slow down the physical, chemical and microbiological degradation.

Yogurt Products

There are two types of plain yogurt:

- Stirred style yogurt
- Set style yogurt.

The above description is essentially the manufacturing proceedures for stirred style. In set style, the yogurt is packaged immediately after inoculation with the starter and is incubated in the packages. Other yogurt products include:

Fruit-on-the-bottom style: fruit mixture is layered at the bottom followed by inoculated yogurt, incubation occurs in the sealed cups.

Soft-serve and Hard Pack Frozen Yogurt

Continental, French, and Swiss: stirred style yogurt with fruit preparation.

Yogurt Beverages

Drinking yogurt is essentially stirred yogurt which has a total solids content not exceeding 11% and which has undergone homogenization to further reduce the viscosity, Flavouring and colouring are invariably added. Heat treatment may be applied to extend the storage life. HTST pasteurization with aseptic processing will give a shelf life of several weeks at 2-4°C, which UHT processes with aseptic packaging will give a shelf life of several weeks at room temperature.

Other Fermented Milk Beverages

Cultured Buttermilk

This product was originally the fermented byproduct of butter manufacture, but today it is more common to produce cultured buttermilks from skim or whole milk. The culture most frequently used in S. lactis,, perhaps also spp. cremoris. Milk is usually heated to 95°C and cooled to 20-25°C before the addition of the starter culture. Starter is added at 1-2% and the fermentation is allowed to proceed for 16-20 hours, to an acidity of 0.9% lactic acid. This product is frequently used as an ingredient in the baking industry, in addition to being packaged for sale in the retail trade.

Acidophilus Milk

Acidophilus milk is a traditional milk fermented with Lactobacillus acidophilus (LA), which has been thought to have therapeutic benefits in the gastrointestinal tract. Skim or whole milk may be used.

The milk is heated to high temperature, e.g., 95°C for 1 hour, to reduce the microbial load and favour the slow growing LA culture. Milk is inoculated at a level of 2-5% and incubated at 37°C until coagulated. Some acidophilus milk has an acidity as high as 1% lactic acid, but for therapeutic purposes 0.6-0.7% is more common.

Another variation has been the introduction of a sweet acidophilus milk, one in which the LA culture has been added but there has been no incubation. It is thought that the culture will reach the GI tract where its therapeutic effects will be realized, but the milk has no fermented qualities, thus delivering the benefits without the high acidity and flavour, considered undesirable by some people.

Sour Cream

Cultured cream usually has a fat content between 12-30%, depending on the required properties. The starter is similar to that used for cultured buttermilk. The cream after standardization is usually heated to 75-80°C and is homogenized at >13 MPa to improve the texture. Inoculation and fermentation conditions are also similar to those for cultured buttermilk, but the fermentation is stopped at an acidity of 0.6%.

Others

There are a great many other fermented dairy products, including kefir, koumiss, beverages based on bulgaricus or bifidus strains, labneh, and a host of others. Many of these have developed in regional areas and, depending on the starter organisms used, have various flavours, textures, and components from the fermentation process, such as gas or ethanol.

Whipped Cream Structure

The structure of whipped cream is very similar to the fat and air structure that exists in ice cream. Cream is an emulsion with a fat content of 35-40%. When you whip a bowl of heavy cream, the agitation and the air bubbles that are added cause the fat globules to begin to partially coalesce in chains and clusters and adsorb to and spread around the air bubbles.

As the fat partially coalesces, it causes one fat-stabilized air bubble to be linked to the next, and so on. The whipped cream soon starts to become stiff and dry appearing and takes on a smooth texture. This results from the formation of this partially coalesced fat structure stabilizing the air bubbles. The water, lactose and proteins are trapped in the spaces around the fat-stabilized air bubbles. The crystalline fat content is essential (hence whipping of cream is very temperature dependent) so that the fat globules partially coalesce into a 3-dimensional structure rather than fully coalesce into larger and larger globules that are not capable of structure-building.

The structure of whipped cream as determined by scanning electron microscopy. A. Overview showing the relative size and prevalence of air bubbles (a) and fat globules (f); bar = 30 um. B. Internal structure of the air bubble, showing the layer of partially coalesced fat which has stabilized the bubble; bar = 5 um. C. Details of the partially coalesced fat layer, showing the interaction of the individual fat globules. Bar = 3 um.

Ice Cream

Ice cream has a long history as a popular dairy food item. It has evolved from a manually manufactured household product to a very automated industrial product. This is the ice cream homepage, a subset of the Dairy Technology Education Series. If you have come to this page directly, then you can go back to the beginning to start learning about dairy science and technology and dairy products.

Ice Cream: History and Folklore

Most of the following material has been extracted from "The History of Ice Cream", written by the International Association of Ice Cream Manufacturers (IAICM), Washington DC, 1978. As you will note below, however, much of the early history of ice cream remains unproven folklore.

Once upon a time, hundreds of years ago, Charles I of England hosted a sumptous state banquet for many of his friends and family. The meal, consisting of many delicacies of the day, had been simply superb but the "coup de grace" was yet to come.

After much preparation, the King's french chef had concocted an apparently new dish. It was cold and resembled fresh-fallen snow but was much creamier and sweeter than any other after-dinner dessert. The guests were delighted, as was Charles, who summoned the cook and asked him not to divulge the recipe for his frozen cream. The King wanted the delicacy to be served only at the Royal table and offered the cook 500 pounds a year to keep it that way. Sometime later, however, poor Charles fell into disfavour with his people and was beheaded in 1649. But by that time, the secret of the frozen cream remained a secret no more. The cook, named DeMirco, had not kept his promise.

This story is just one of many of the fascinating tales which surround the evolution of our country's most popular dessert, ice cream. It is likely that ice cream was not invented, but rather came to be over years of similar efforts. Indeed, the Roman Emperor Nero Claudius Caesar is said to have sent slaves to the mountains to bring snow and ice to cool and freeze the fruit drinks he was so fond of. Centuries later, the Italian Marco Polo returned from his famous journey to the Far East with a recipe for making water ices resembling modern day sherbets.

Most books are full of myths about the history of ice cream. According to popular accounts, Marco Polo (1254-1324) saw ice creams being made during his trip to China, and on his return, introduced them to Italy. The myth continues with the Italian chefs of the you Catherine de'Medici taking this magical dish to France when she went there in 1533 to marry the Duc d'Orleans, with Charles I rewarding his own ice-cream maker with a lifetime pension on condition that he did not divulge his secret recipe to anyone, thereby keeping ice cream as a royal perogative.

Unfortunately, there is no historical evidence to support any of these stories. They would appear to be purely the creation of imaginative nineteenth-century ice-cream makers and vendors. Indeed, we have found no mention of any of these stories before the nineteenth century. They go on to refute the claims about Marco Polo, Catherine de'Medici, and Charles I (in particular, while the IAICM reference credits DeMirco as the Charles I chef, apparently while other various sources credit 10 different men, there are no records of such a pension being paid to any of Charles I's cooks). They do go on in their book to discuss history for which there is a record, with (I think) the earliest written record being something made in China.

Chris Clarke, in his 2004 Royal Society of Chemistry mongraph "The Science of Ice Cream", points out quite correctly that the history of ice cream is closely associated with the development of refrigeration techniques and can thus be traced in several stages:

1. Cooling food and rink by mixing it with snow or ice;
2. The discosvery that dissolving salts in water produces cooling;
3. The discovery (and spread of knowledge) that mixing salts and snow or ice cools even further-mid to late 17th century-the inclusion of cream in the water ices also evolved around this time;
4. The invention of the ice cream maker in the mid 19th century;
5. The development of mechanical refrigeration in the later 19th and early 20th centuries-which led to the development of the modern ice cream industry.

Back to the IAICM History

In 1774, a caterer named Phillip Lenzi announced in a New York newspaper that he had just arrived from London and would be offering for sale various confections, including ice cream. Dolley Madison, wife of U.S. President James Madison, served ice cream at her husband's Inaugural Ball in 1813.

The first improvement in the manufacture of ice cream (from the handmade way in a large bowl) was given to us by a New Jersey woman, Nancy johnson, who in 1846 invented the hand-cranked freezer. This device is still familiar to many.

By turning the freezer handle, they agitated a container of ice cream mix in a bed of salt and ice until the mix was frozen. Because Nancy Johnson lacked the foresight to have her invention patented, her name does not appear on the patent records. A similar type of freezer was, however, patented on May 30, 1848, by a Mr. Young who at least had the courtesy to call it the "Johnson Patent Ice Cream Freezer".

Ice Cream Formulations

Ice Cream Mix General Composition:

- Milkfat: >10%-16%
- Milk solids-not-fat (snf): 9%-12%
- Sucrose: 10%-14%
- Corn syrup solids: 4%-5%
- Stabilizers: 0%-0.4%

- Emulsifiers: 0%-0.25%
- Water: 55%-64%.

The snf contains, on average, dry wt. basis, 38% protein, 54% lactose, and 8% ash (including 1.38% Ca, 1.07% P, 1.22% K, 0.7% Na).

Formulation Considerations

First of all, regulatory issues. What are the product definitions for your legal jurisdiction? These, of course, have to be met.

Next, desired Fat and Total solids (%):-Quality considerations, what kind of product are you trying to make?

Fat: MSNF Ratio's usually determined next based on fat content.

Sugar: Glucose solids Ratio's determined based on fat and total solids requirements, sweetness, freezing point depression, body and shelf life desired, and cost considerations.

Stabilizer/Emulsifier considerations come last, based on the ice cream formulation, and processing and distribution factors involved in each application.

With these considerations in mind, it is also useful to look at ranges of components for the categories of products in the *ice cream* category that are available in the market.

These are industry-used terms, not legally defined.

Economy Brands

- Fat content, usually legal minimum, e.g., 10%
- Total solids, usually legal minimum, e.g., 36%
- Overrun, usually legal maximum, ~120%
- Cost, low.

Standard Brands

- Fat content, 10-12%
- Total solids, 36-38%
- Overrun, 100-120%
- Cost, average.

Premium Brands

- Fat content, 12-15%
- Total solids, 38-40%
- Overrun, 60-90%
- Cost, higher than average.

Super-premium Brands

- Fat content, 15-18%
- Total solids, >40%
- Overrun, 25-50%
- Cost, high.

Suggested Mixes

Suggested mixes for hard-frozen ice cream products.

	Percent (%)						
Milk Fat	10.0	11.0	12.0	13.0	14.0	15.0	16.0
Milk Solids-not-fat	11.0	11.0	10.5	10.5	10.0	10.0	9.5
Sucrose	10.0	10.0	12.0	14.0	14.0	15.0	15.0
Corn Syrup Solids	5.0	5.0	4.0	3.0	3.0	-	-
Stabilizer*	0.35	0.35	0.30	0.30	0.25	0.20	0.15
Emulsifier*	0.15	0.15	0.15	0.14	0.13	0.12	0.10
Total Solids	36.5	37.5	38.95	40.94	41.38	40.32	40.75

- Highly variable depending on type; manufacturers recommendations are usually followed.
- Usually an inverse relationship between fat and total solids compared to snf
- Generally an inverse relationship between glucose solids (corn sweetener) levels and total solids
- As total solids increases, there is less requirement for stabilizer
- As fat levels in a mix increase, there is generally less need for emulsifier.

Suggested mixes for low-fat (3-5% fat) and light (6-8% fat) ice cream products.

	Percent (%)				
Milk Fat	3.0	4.0	5.0	6.0	8.0
Milk SNF	13.0	12.5	12.5	12.0	11.5
Sucrose	11.0	11.0	11.0	13.0	12.0
CSS	6.0	5.5	5.5	4.0	4.0
Stabilizer	0.35	0.35	0.35	0.35	0.35
Emulsifier	0.10	0.10	0.10	0.15	0.15
Total Solids	33.65	33.45	34.45	35.5	36.0

Ice milk was the traditional lower fat ice cream product for many years, but this category has been re-classified by many regulatory jurisdictions to include three reduced fat categories: light ice cream, lowfat ice cream (the traditional ice milk), and non-fat ice cream. It has generally been possible to produce fat contents as low as 4% with traditional products, but further fat reductions have generally involved fat-replacers.

Table: *Suggested mixes for soft-frozen ice cream products*

	Percent (%)	
Milk Fat	10.0	10.0
Milk Solids-not-fat	12.5	12.0
Sucrose	13.0	10.0
Corn Syrup Solids	—-	4.0
Stabilizer*	0.35	0.15
Emulisifier*	0.15	0.15
Total Solids	36.0	36.3

*Highly variable depending on type; manufacturers recommendations are usually followed.

- Generally, while the fat content is kept lower, the snf content is generally higher than for hard-frozen products.
- Glucose solids are.often used, but can lead to an enhanced sensation of guminess.
- Stabilizers are also generally used for viscosity enhancement and mouthfeel, but their function in ice recrystallization is no longer needed.
- Dryness, however, is a big concern in soft-serve products, hence the emulsifier content is generally kept high.

Sherbet and Sorbet

	Percent (%)	
Milk Fat	0.5	1.5
Milk Solids-not-fat	2.0	3.5
Sucrose	24.0	24.0
Corn Syrup Solids	9.0	6.0
Stabilizer/emulsifier*	0.3	0.3
Citric acid (50% sol.)**	0.7	0.7
Water	63.5	64.0
Total	100.0	100.0

* Or as advised from the supplier.

** Acid is added just before freezing, after aging of the mix.

- Sorbet: delete the mix and skim powder
- Fruit: at about 25% to the mix.

Frozen Yogurt

	Percent (%)
Fat	2.0
MSNF	14.0
Sugar	15.0
Stabilizer	0.35
Water	68.65
Total	100.0

- Example Processing Instructions: 20% of this mix, consisting of skimmilk and skimmilk powder blended to give 12.5% solids, is to be incubated as the yogurt portion.
- To make the "incubated" portion, combine the appropriate amount of skimmilk and skimmilk powder, pasteurize at a high temperature, cool to 104 to 110F, and inoculate with a yogurt culture (typical of yogurt processing). When the fermentation is complete (to the desired acidity), cool the "yogurt".
- To make the "sweet" mix, combine the cream, sugar and stabilizer, and the balance of the skimmilk powder and skimmilk, pasteurize, homogenize, cool (typical for ice cream processing), and blend with the "yogurt".
- The completed frozen yogurt mix is then aged and prepared for flavouring and freezing.

Ice Cream Ingredients

Ice cream has the following composition:

- greater than 10% milkfat by legal definition, and usually between 10% and as high as 16% fat in some premium ice creams
- 9 to 12% milk solids-not-fat: this component, also known as the serum solids, contains the proteins (caseins and whey proteins) and carbohydrates (lactose) found in milk
- 12 to 16% sweeteners: usually a combination of sucrose and glucose-based corn syrup sweeteners
- 0.2 to 0.5% stabilizers and emulsifiers
- 55% to 64% water which comes from the milk or other ingredients

These percentages are by weight, either in the mix or in the frozen ice cream. Please remember, however, that when frozen, about one half of the volume of ice cream is air, so by volume in ice cream, these numbers can be reduced by approximately one-half, depending on the actual air content.

However, since air does not contribute weight, we usually talk about the composition of ice cream on a weight basis, bearing in mind this important distinction. All ice cream flavours, with the possible exception of chocolate, are made from a basic white mix.

Formulations can be derived from a number of different starting points. Details and suggested formulas are detailed on the formulations page, but turning the formulation into a recipe depends on the ingredients used to supply the components, and it is then necessary to do a mix calculation to determine the required ingredients based on the formula. Ice milk and light ice creams are very similar to the composition of ice cream but in the case of ice milk in Canada, for example, it must contain between 3% and 5% milkfat by legal definition.

The ingredients to supply the desired components are chosen on the basis of availability, cost, and desired quality.

Milkfat (or "Butterfat") / Fat

Milkfat, or fat in general, including that from non-dairy sources, is important to ice cream for the following reasons:

- increases the richness of flavour in ice cream
- produces a characteristic smooth texture by lubricating the palate
- helps to give body to the ice cream, due to its role in fat destabilization
- aids in good melting properties, also due to its role in fat destabilization
- aids in lubricating the freezer barrel during manufacturing (Non-fat mixes are extremely hard on the freezing equipment).

The limitations of excessive use of butterfat in a mix include:

- cost
- hindered whipping ability
- decreased consumption due to excessive richness
- high caloric value.

The best source of butterfat in ice cream for high quality flavour and convenience is fresh sweet cream from fresh sweet milk. Other sources include butter or anhydrous milkfat.

During freezing of ice cream, the fat emulsion which exists in the mix will partially destabilize or churn as a result of the air incorporation, ice crystallization and high shear forces of the blades. This partial churning is necessary to set up the structure and texture in ice cream, which is very similar to the structure in whipped cream. Emulsifiers help to promote this destabilization process. The triglycerides in milkfat have a wide melting range, +40° C to -40° C, and thus there is always a combination of liquid and crystalline fat. Alteration of this solid: liquid ratio can affect the amount of fat destabilization that occurs. Duplicating this structure with other sources of fat is difficult.

Vegetable (non-dairy) fats are used extensively as fat sources in ice cream in the United Kingdom, parts of Europe, the Far East, and Latin America but only to a very limited extent in North America. Five factors of great interest in selection of fat source are the crystal structure of the fat, the rate at which the fat crystallizes during dynamic temperature conditions, the temperature-dependent melting profile of the fat, especially at chilled and freezer temperatures, the content of high melting triglycerides (which can produce a waxy, greasy mouthfeel) and the flavour and purity of the oil.

It is important that the fat droplet contain an intermediate ratio of liquid:solid fat at the time of freezing. It is difficult to quantify this ratio as it is dependent on a number of composition and manufacturing factors, however, 1/2 to 2/3 crystalline fat at 4-5°C is a good, working rule. Crystallization of fat occurs in three steps: undercooling to induce nucleation, heterogeneous or homogeneous nucleation (or both), and crystal propagation.

In bulk fat, nucleation is predominantly heterogeneous, with crystals themselves acting as nucleating agents for further crystallization, and undercooling is usually minimal. However, in an emulsion, each droplet must crystallize independently of the next. For heterogeneous nucleation to predominate, there must be a nucleating agent available in every droplet, which is often not the case. Thus in emulsions, homogeneous nucleation and extensive undercooling may be common. Blends of oils are often used in ice cream manufacture, selected to take into account physical characteristics, flavour, availability, stability during storage and cost.

We have recently completed a study on the use of non-dairy fats in frozen desserts, which is available here. A blend of 75% of either fractionated palm kernel oil or coconut oil and 25% of an unsaturated oil, like high oleic sunflower oil, was shown to produce optimal levels of fat destabilization, meltdown and flavour, although coconut oil may take longer to crystallize during aging. Blends of 50% milkfat, 37.5% fractionated palm kernel or coconut oil, and 12.5% high oleic sunflower oil were also shown to be very acceptable.

Milk Solids-not-fat

The serum solids or milk solids-not-fat (MSNF) contain the lactose, caseins, whey proteins, minerals, and ash content of the product from which they were derived. They are an important ingredient for the following beneficialreasons:

- improve the texture of ice cream, due to the protein functionality
- help to give body and chew resistance to the finished product
- are capable of allowing a higher overrun without the characteristic snowy or flaky textures associated with high overrun, due also to the protein functionality
- may be a cheap source of total solids, especially whey powder

The limitations on their use include off flavours which may arise from some of the products, and an excess of lactose which can lead to the defect of sandiness prevelant when the lactose crystallizes out of solution. Excessive concentrations of lactose in the serum phase may also lower the freezing point of the finished product to an unacceptable level.

The best sources of serum solids for high quality products are:

- concentrated skimmed milk
- spray process low heat skimmilk powder.

Other sources of serum solids include: sweetened condensed whole or skimmed milk, frozen condensed skimmed milk, buttermilk powder or condensed buttermilk, condensed whole milk, or dried or condensed whey. Superheated condensed skimmed milk, in which high viscosity is promoted, is sometimes used as a stabilizing agent but does, then, also contribute to serum solids.

It has recently become common practice to replace the use of skim milk powder or condensed skim with a variety of milk powder replacers, which are blends of whey protein concentrates, caseinates, and whey powders. These are formulated with less protein than skim powder, usually 20-25% protein, and thus less cost, but are blended with an appropriate balance of whey proteins and caseins to do an adequate job. Caution must be exercised in excessive use of these powders, experimentation with your own mix is the best answer.

See the section on Concentrated and Dried Dairy Products for a description of the manufacture of all of the above ingredients.

The proteins, which make up approximately 4% of the mix, contribute much to the development of structure in ice cream including:

- emulsification properties in the mix
- whipping properties in the ice cream
- water holding capacity leading to enhanced viscosity and reduced iciness.

Lactose Crystallization

1. A decrease in temperature favours rapid crystallization insofar as it increases the supersaturation.
2. A decrease in temperature favours slow crystallization insofar as it increases the viscosity, reduces the kinetic energy of the particles, and decreases the rate of transformation from beta to alpha lactose.

Supersaturated state can exist, however, due to extreme viscosity, and it is likely that much of the lactose in ice cream is non-crystalline. Stabilizers help to hold lactose in supersaturated state due to viscosity enhancement. Fruits, nuts, candy-add crystal centres and may enhance lactose crystallization. Nuts pull out moisture from ice cream immediately surrounding the nut thus concentrating the mix. Citrate and phosphate ions decrease tendency for fat coalescence (Sodium citrate, Disodium Phosphate). They prevent churning in soft ice cream for example, producing a wetter product. These salts decrease the degree of protein aggregation. Calcium and magnesium ions have the opposite effect, promote partial coalescence. Calcium sulfate, for example, results in a drier ice cream. Calcium and Magnesium increase the degree of protein aggregation.

Salts may also influence electrostatic interactions. Fat globules carry a small net negative charge, these ions could increase or decrease that charge as they were attracted to or repelled from surface.

Sweeteners

A sweet ice cream is usually desired by the consumer. As a result, sweetening agents are added to ice cream mix at a rate of usually 12-16% by weight. Sweeteners improve the texture and palatability of the ice cream, enhance flavours, and are usually the cheapest source of total solids.

In addition, the sugars, including the lactose from the milk components, contribute to a *depressed freezing point* so that the ice cream has some unfrozen water associated with it at very low temperatures typical of their serving temperatures, -15° to -18° C. Without this unfrozen water, the ice cream would be too hard to scoop. The effect of sweeteners on freezing characteristics of ice cream mixes is demonstrated by the plot shown on the ice cream freezing curve.

Sucrose is the main sweetener used because it imparts excellent flavour. Sucrose is a disaccharide made up of glucose (dextrose, cerelose), and fructose (levulose). Sucrose is dextrorotatory-meaning it rotates a plane of polarized light to the right, + 66.5°. With hydrolyzed sucrose the plane of polarization is to the left, "inverted" -20°. An acid, plus water, plus heat treatment, at concentrations above 10%, yields invert sugar and increases the sweetness. It has become common in the industry to substitute all or a portion of the sucrose content with sweeteners derived from corn syrup. This sweetener is reported to contribute a firmer and more chewy body to the ice cream, is an economical source of solids, and improves the shelf life of the finished product.

Corn syrup in either its liquid or dry form is available in varying dextrose equivalents (DE). The DE is a measure of the reducing sugar content of the syrup calculated as dextrose and expressed as a percentage of the total dry weight. As the DE is increased by hydrolysis of the corn starch, the sweetness of the solids is increased and the average molecular weight is decreased. This results in an increase in the freezing point depression, in such foods as ice cream, by the sweetener. The lower DE corn syrup contains more dextrins which tie up more water in the mix thus supplying greater stabilizing effect against coarse texture.

An enzymatic hydrolysis and isomerization procedure can convert glucose to fructose, a sweeter carbohydrate, in corn syrups thus producing a blend (high fructose corn syrup, HFCS) which can be used to a much greater extent in sucrose replacement. However, these HFCS blends further reduce the freezing point producing a very soft ice cream at usual conditions of storage and dipping in the home.

Here is a diagram illustrating the effect of DE and maltose or fructose conversion on the properties of corn starch hydrolysates as used in ice cream. A balance is involved between sweetness, total solids, and freezing point.

Stabilizers

The stabilizers are a group of compounds, usually polysaccharide food gums, that are responsible for adding viscosity to the mix and the unfrozen phase of the ice cream. This results in many functional benefits, listed below, and also extends the shelf life by limiting ice recrystallization during storage. Without the stabilizers, the ice cream would become coarse and icy very quickly due to the migration of free water and the growth of existing ice crystals.

The smaller the ice crystals in the ice cream, the less detectable they are to the tongue. Especially in the distribution channels of today's marketplace, the supermarkets, the trunks of cars, and so on, ice cream has many opportunities to warm up, partially melt some of the ice, and then refreeze as the temperature is once again lowered. This process is known as heat shock and every time it happens, the ice cream becomes more icy tasting. Stabilizers help to prevent this.

The functions of stabilizers in ice cream are:

- In the mix: To stabilize the emulsion to prevent creaming of fat and, in the case of carrageenan, to prevent serum separation due to incompatibility of the other polysaccharides with milk proteins, also to aid in suspension of liquid flavours
- In the ice cream at draw from the scraped surface freezer: To stabilize the air bubbles and to hold the flavourings, e.g., ripple sauces, in dispersion
- In the ice cream during storage: To prevent lactose crystal growth and retard or reduce ice crystal growth during storage, also to prevent shrinkage from collapse of the air bubbles and to prevent moisture migration into the package (in the case of paperboard) and sublimation from the surface
- In the ice cream at the time of consumption: To provide some body and mouthfeel without being gummy, and to promote good flavour release.

Limitations on their use include:

- production of undesirable melting characteristics, due to too high viscosity
- excessive mix viscosity prior to freezing
- contribution to a heavy or chewy body.

The stabilizers in use today include:

Locust Bean Gum : soluble fibre of plant material derived from the endosperm of beans of exotic trees grown mostly in Africa (Note: locust bean gum is a synonym for carob bean gum, the beans of which were used centuries ago for weighing precious metals, a system still in use today, the word carob and Karat having similar derivation).

Guar Gum: from the endosperm of the bean of the guar bush, a member of the legume family grown in India for centuries and now grown to a limited extent in Texas.

Carboxymethyl Cellulose (CMC): derived from the bulky components, or pulp cellulose, of plant material, and chemically derivatized to make it water soluble.

Xanthan Gum: produced in culture broth media by the microorganism *Xanthaomonas campestris* as an exopolysaccharide, used to a lesser extent.

Sodium Alginate: an extract of seaweed, brown kelp, also used to a lesser extent.

Carrageenan: an extract of Irish Moss or other red algae, originally harvested from the coast of Ireland, near the village of Carragheen but now most frequently obtained from Chile and the Phillipines.

Each of the stabilizers has its own characteristics and often, two or more of these stabilizers are used in combination to lend synergistic properties to each other and improve their overall effectiveness. Guar, for example, is more soluble than locust bean gum at cold temperatures, thus it finds more application in HTST pasteurization systems. Carrageenan is not used by itself but rather is used as a secondary colloid to prevent the wheying off of mix which is usually promoted by one of the other stabilizers.

Gelatin, a protein of animal origin, was used almost exclusively in the ice cream industry as a stabilizer but has gradually been replaced with polysaccharides of plant origin due to their increased effectiveness and reduced cost.

Emulsifiers

The emulsifiers are a group of compounds in ice cream that aid in developing the appropriate fat structure and air distribution necessary for the smooth eating and good meltdown characteristics desired in ice cream. Since each molecule of an emulsifier contains a hydrophilic portion and a hydrophobic portion, they reside at the interface between fat and water. As a result they act to reduce the interfacial tension or the force which exists between the two phases of the *emulsion*.

This causes a desorption of protein from the fat droplet surface, which promotes a destabilization of the fat emulsion (due to a weaker membrane) leading to a smooth, dry product with good meltdown properties. Their action will be more fully explained in the structure of ice cream section.

The original ice cream emulsifier was egg yolk, which was used in most of the original recipes. Today, two emulsifiers predominate most ice cream formulations:

Mono-and di-glycerides: derived from the partial hydrolysis of fats or oils of animal or vegetable origin.

Polysorbate 80: a sorbitan ester consisting of a glucose alcohol (sorbitol) molecule bound to a fatty acid, oleic acid, with oxyethylene groups added for further water solubility.

Other possible sources of emulsifiers include buttermilk, and glycerol esters. All of these compounds are either fats or carbohydrates, important components in most of the foods we eat and need. Together, the stabilizers and emulsifiers make up less than one half percent by weight of our ice cream. They are all compounds which have been exhaustively tested for safety and have received the "generally recognized as safe" or GRAS status.

Mix Calculations for Ice Cream and Frozen Dairy Desserts

The general objective in calculating ice cream mixes is to turn your formula into a recipe based on the ingredients you intend to use and the amount of mix you desire.

The formula is given as percentages of fat, milk solids-not-fat, sugar, corn syrup solids (glucose solids), stabilizers and emulsifiers. The ingredients to supply these components are chosen on the basis of availability, quality and cost.

The following table illustrates the relationship between the major components, the main ingredients that supply the major components, and the minor components that are supplied with the major ones for each ingredient.

Component and Ingredients to supply that component (but note that each of these ingredients also supplies the following other components): Milkfat, supplied by Cream (which also supplies SNF and water) or Butter (which also supplies SNF, water); Milk solids-not-fat (SNF, or sometimes also called serum solids, S.S.), supplied by any of the following:

- Skim powder (which also supplies water, about 3%)
- Condensed skim (which also supplies water)
- Condensed milk (which also supplies water and fat)
- Sweetened condensed (which also supplies water and sugar)
- Whey powder (which also supplies water).

Water, supplied by Skim milk (which also supplies msnf), or milk (which also supplies fat and msnf), or pure water.

Sweetener, supplied by dry or liquid (which also then supplies water) sucrose or corn syrup solids. The first step in a mix calculation is to identify for each ingredient we intend to use its components. If there is only one source of the component we need for the formula, for example the stabilizer or the sugar, we determine it directly by multiplying the percentage we need by the amount we need, e.g., 100 kg of mix @ 10% sugar would require 10 kg sugar. If there are two or more sources, for example we need 10 % fat and it is coming from both cream and milk, then we need to utilize an algebraic method.

Computer programs developed for mix calculations generally solve a simultaneous equation based on mass and component balances. To solve simultaneous equations, you need as many independent equations as you have unknowns.

For manual calculations, a method known as the "Serum Point" method has been derived. This method has solved the simultaneous equations in a general way so that only the equations need to be known and not resolved each time.

In standardizing mixes, the composition of the various ingredients used must be known. In some cases the percentage of solids contained in a product is taken as constant, while in others the composition must be obtained by analysis. Information on the various ingredients is given below:

(a) Skim milk-can be determined by analysis or assumed at 9 percent serum solids. Fat (0.01%-0.10%) should be taken into account if significant.

(b) Dried products, e.g. skim milk powder, whey powder, WPC, milk powder blends, usually taken to be 97 percent solids as they retain some moisture.

(c) Cream-Percent fat usually measured by an acceptable method.

Percent MSNF found by formula as follows: (100-percent fat) x.09 = % snf (assuming that the "skim milk" contains 9% total solids). Example: In cream testing 30% fat, the percent snf would be (100-30) x.09 = 6.3% snf

(d) Milk-Percent fat measured by an acceptable method.

Percent snf may be found same as for cream or by making a total solids test and deducting the percent fat.

(e) Condensed Milk Products-Composition of these products should be obtained by the supplier.

(f) Sweeteners-Sucrose-Dry 100% solids

Sucrose-Liquid 66% solids

Dextrose-Dry 100% solids

Corn Syrup Solids 100% solids

Corn Syrup Liquid 80% solids

Glucose 80% solids

Honey 80% solids

(g) Stabilizers and Emulsifiers (if solid)-Because of the small percentage used may be figured as 100 percent solids.

(h) Egg Products-Fresh whole eggs: 10% fat, 25% solids

Fresh egg yolk: 33% fat, 50% solids

Frozen egg yolk: 33% fat, 50% solids

Dried egg yolk: 60% fat, 100% solids

Below are some example problems to look at, if you are interested in the mathematics of mix calculations.

- Example 1. Basic mix using butter, skim powder, and water (only one source of each component). (Algebraic Method)
- Example 2. Mix using cream, skim, and skim powder (three sources of milk SNF, three sources of water). (Algebraic and Serum Point Methods)
- Example 3. Mix using cream, milk, and skim powder (three sources of milk SNF, three sources of water, and two source of fat). (Algebraic and Serum Point Methods)
- Example 4. Mix using cream, milk, and sweetened, condensed skim (three sources of milk SNF, three sources of water, two sources of fat, and two sources of sugar). (Serum Point Method)
- Example 5. Mix using cream, milk, and sweetened, condensed milk (three sources of milk SNF, three sources of water, three sources of fat, and two sources of sugar). (Serum Point Method)
- Example 6. Mix using a given amount of cream and skim, with the balance coming from butter, milk, and skim powder. (Serum Point Method)
- Example 7. Mix using cream, milk, condensed skim, and liquid sweeteners (water needs to be accounted for). (Serum Point Method)

After completing a problem, you should do a proof of your calculation, by ensuring that the mass sums to the desired value, and that the mass fraction of all components also sum to the desired value. There is only one unique solution, so you know by calculation if you have it right or not!

Ice Cream Manufacture

The basic steps in the manufacturing of ice cream are generally as follows:

- blending of the mix ingredients
- pasteurization
- homogenization
- aging the mix
- freezing
- packaging
- hardening.

Process flow diagram for ice cream manufacture: the red section represents the operations involving raw, unpasteurized mix, the pale blue section represents the operations involving pasteurized mix, and the dark blue section represents the operations involving frozen ice cream.

Blending

First the ingredients are selected based on the desired formulation and the calculation of the recipe from the formulation and the ingredients chosen, then the ingredients are weighed and blended together to produce what is known as the "ice cream mix". Blending requires rapid agitation to incorporate powders, and often high speed blenders are used.

Pasteurization

The mix is then pasteurized. Pasteurization is the biological control point in the system, designed for the destruction of pathogenic bacteria. In addition to this very important function, pasteurization also reduces the number of spoilage organisms such as psychrotrophs, and helps to hydrate some of the components (proteins, stabilizers).

Batch pasteurizers lead to more whey protein denaturation, which some people feel gives a better body to the ice cream. In a batch pasteurization system, blending of the proper ingredient amounts is done in large jacketed vats equipped with some means of heating, usually steam or hot water. The product is then heated in the vat to at least 69 C (155 F) and held for 30 minutes to satisfy legal requirements for pasteurization, necessary for the destruction of pathogenic bacteria.

Various time temperature combinations can be used. The heat treatment must be severe enough to ensure destruction of pathogens and to reduce the bacterial count to a maximum of 100,000 per gram. Following pasteurization, the mix is homogenized by means of high pressures and then is passed across some type of heat exchanger (plate or double or triple tube) for the purpose of cooling the mix to refrigerated temperatures (4 C). Batch tanks are usually operated in tandem so that one is holding while the other is being prepared. Automatic timers and valves ensure the proper holding time has been met.

Continuous pasteurization is usually performed in a high temperature short time (HTST) heat exchanger following blending of ingredients in a large, insulated feed tank. Some preheating, to 30 to 40 C, is necessary for solubilization of the components. The HTST system is equipped with a heating section, a cooling section, and a regeneration section. Cooling sections of ice cream mix HTST presses are usually larger than milk HTST presses. Due to the preheating of the mix, regeneration is lost and mix entering the cooling section is still quite warm.

Homogenization

The mix is also homogenized which forms the fat emulsion by breaking down or reducing the size of the fat globules found in milk or cream to less than 1 μ m. Two stage homogenization is usually preferred for ice cream mix. Clumping or clustering of the fat is reduced thereby producing a thinner, more rapidly whipped mix. Melt-down is also improved. Homogenization provides the following functions in ice cream manufacture:

- Reduces size of fat globules
- Increases surface area
- Forms membrane
- makes possible the use of butter, frozen cream, etc.

By helping to form the fat structure, it also has the following indirect effects:

- makes a smoother ice cream
- gives a greater apparent richness and palatability
- better air stability
- increases resistance to melting.

Homogenization of the mix should take place at the pasteurizing temperature. The high temperature produces more efficient breaking up of the fat globules at any given pressure and also reduces fat clumping and the tendency to thick, heavy bodied mixes. No one pressure can be recommended that will give satisfactory results under all conditions. The higher the fat and total solids in the mix, the lower the pressure should be. If a two stage homogenizer is used, a pressure of 2000-2500 psi on the first stage and 500-1000 psi on the second stage should be satisfactory under most conditions. Two stage homogenization is usually preferred for ice cream mix. Clumping or clustering of the fat is reduced thereby producing a thinner, more rapidly whipped mix. Melt-down is also improved.

Ageing

The mix is then aged for at least four hours and usually overnight. This allows time for the fat to cool down and crystallize, and for the proteins and polysaccharides to fully hydrate. Aging provides the following functions:

- Improves whipping qualities of mix and body and texture of ice cream.

It does so by:

- providing time for fat crystallization, so the fat can partially coalesce;
- allowing time for full protein and stabilizer hydration and a resulting slight viscosity increase;
- allowing time for membrane rearrangement and protein/ emulsifier interaction, as emulsifiers displace proteins from the fat globule surface, which allows for a reduction in stabilization of the fat globules and enhanced partial coalescence.

Aging is performed in insulated or refrigerated storage tanks, silos, etc. Mix temperature should be maintained as low as possible without freezing, at or below 5 C. An aging time of overnight is likely to give best results under average plant conditions. A "green" or unaged mix is usually quickly detected at the freezer.

Freezing and Hardening

Following mix processing, the mix is drawn into a flavour tank where any liquid flavours, fruit purees, or colours are added. The mix then enters the dynamic freezing process which both freezes a portion of the water and whips air into the frozen mix. The "barrel" freezer is a scraped-surface, tubular heat exchanger, which is jacketed with a boiling refrigerant such as ammonia or freon. Mix is pumped through this freezer and is drawn off the other end in a matter of 30 seconds, (or 10 to 15 minutes in the case of batch freezers) with about 50% of its water frozen. There are rotating blades inside the barrel that keep the ice scraped off the surface of the freezer and also dashers inside the machine which help to whip the mix and incorporate air.

Ice cream contains a considerable quantity of air, up to half of its volume. This gives the product its characteristic lightness. Without air, ice cream would be similar to a frozen ice cube. The air content is termed its overrun, which can be calculated mathematically.

As the ice cream is drawn with about half of its water frozen, particulate matter such as fruits, nuts, candy, cookies, or whatever you like, is added to the semi-frozen slurry which has a consistency similar to soft-serve ice cream. In fact, almost the only thing which differentiates hard frozen ice cream from soft-serve, is the fact that soft serve is drawn into cones at this point in the process rather than into packages for subsequent hardening.

Hardening

After the particulates have been added, the ice cream is packaged and is placed into a blast freezer at-30° to-40° C where most of the remainder of the water is frozen. Below about -25° C, ice cream is stable for indefinite periods without danger of ice crystal growth; however, above this temperature, ice crystal growth is possible and the rate of crystal growth is dependant upon the temperature of storage. This limits the shelf life of the ice cream. A primer on the theoretical aspects of freezing will help you to fully understand the freezing and recrystallization process.

Hardening invloves static (still, quiescent) freezing of the packaged products in blast freezers. Freezing rate must still be rapid, so freezing techniques involve low temperature (-40oC) with either enhanced convection (freezing tunnels with forced air fans) or enhanced conduction (plate freezers).

The rate of heat transfer in a frezing porcess is affected by the temperature difference, the surface area exposed and the heat transfer coefficient (Q=U A dT). Thus, the factors affecting hardening are those affecting this rate of heat transfer:

- Temperature of blast freezer-the colder the temperature, the faster the hardening, the smoother the product.
- Rapid circulation of air-increases convective heat transfer.
- Temperature of ice cream when placed in the hardening freezer-the colder the ice cream at draw, the faster the hardening;-must get through packaging operations fast.
- Size of container-exposure of maximum surface area to cold air, especially important to consider shrink wrapped bundles-they become a much larger mass to freeze. Bundling should be done after hardening.
- Composition of ice cream-related to freezing point depression and the temperature required to ensure a significantly high ice phase volume.

- Method of stacking containers or bundles to allow air circulation. Circulation should not be impeded-there should be no 'dead air' spaces (e.g., round vs. square packages).
- Care of evaporator-freedom from frost-acts as insulator.
- Package type, should not impede heat transfer-e.g., styrofoam liner or corrugated cardboard may protect against heat shock after hardening, but reduces heat transfer during freezing so not feasible.

Ice Cream Novelty/Impulse Products

Moulded Novelties

Ice cream novelties are typically single-serving items bought from vendors for immediate consumption. They come in a variety of shapes, sizes, colours, flavours, etc., and manufacturers are constantly developing new items to compete for market share. hence, this category is variably known as either impulse products or novelty products. These products can be manufactured by a number of processes. Products such as popsicles or ice cream with flat sides on sticks are manufactured by filling a mold with water ice or soft ice cream from a barrel freezer, immersion freezing that mold in a cold bath of calcim chloride solution, inserting the stick when partially hardened, and then after hardening is complete, removal of the ice cream from the mold by the stick.

Point 1 = filling;

Point 2 = partial freezing as the molds progress in the calcium chloride bath-you can have an optional "suck-out" of unfrozen material and refillng of something else at this stage;

Point 3 = stick insertion-the product is now frozen sufficiently to hold the stick in place;

Point 4 = further freezing as the molds progress in the calcium chloride bath-you can insert ribbon sauces anywhere along this stage, until the product gets too hard;

Point 5 = withdrawal of the mold from the calcium chloride bath, a quick defrost to free the bar from the mold wall, and extraction of the product from the mold by grabbing the stick-after which the product can be coated with nuts, enrobed with chocolate, then packaged, and on to storage. The molds continue underneath the machinewhere they are washed, rinsed, and sanitized, ready for the next filling.

Extruded Novelties

Products with no sticks, like bars with fancy shapes, or with sticks but irregular sides, such that they could not be pulled out of a mold, are frozen by extrusion. In such a process, the shape is formed and sliced, either vertically or horizontally as below, that slice is further hardenend by passing it through a cold tunnel, and then nuts or syrup can be coated on it, it can be enrobed with chocolate sauce, or whatever is desired to give the final product.

Horizontal extrusion is used to make the "chocolate bar analogue" type of products, while vertical extrusion is used to make everythng from chocolate-enrobed ice cream slices to fancy-shaped novelties on a stick.

Ice Cream Flavours

Most ice cream is purchased by the consumer on basis of flavour and ingredients. There are many different flavours of ice cream manufactured, and to some extent limited only by imagination. Vanilla accounts for 30% of the ice cream consumed.

This is partly because it is used in so many products, like milkshakes, sundaes, banana splits, in addition to being consumed with pies, desserts, etc. It is the ice cream manufacturers responsibility to prepare an excellent mix, but often they put the responsibility of the flavours and ingredients on the supplier.

Ingredients are added to ice cream in four ways during the manufacturing process:

1. Mix Tank: for liquid flavours, colours, fruit purees, flavoured syrup bases Đ anything that will be homogeneously distributed in the frozen ice cream.
2. Variegating Pump: for ribbons, swirls, ripples, revels
3. Ingredient Feeder: for particulates-fruits, nuts, candy pieces, cookies, etc., some complex flavours may utilize 2 feeders
4. Shaker table: for large inclusions.

Generally, the delicate, mild flavours are easily blended and tend not to become objectionable at high concentrations, while harsh flavours are usually objectionable even in low concentrations. Therefore, delicate flavours are preferable to harsh flavours, but in any case a flavour should only be intense enough to be easily recognized. Flavouring materials may be:

1. Natural
2. Artificial or imitation
3. Blends of the two.

***Table:** US Ice Cream Consumption by Flavour, 2006*

Type of Ice-cream	*Percentage of Volume*
1. Vanilla	30.2
2. Chocolate	10.0
3. Chocolate Chip	5.7
4. Butter Pecan	4.0
5. Strawberry	3.7
6. Neapolitan	3.0
7. Cookies and cream	2.6
8. Rocky Road	1.9
9. Cookie Dough	1.5
10. Cherry Vanilla	0.9
11. Coffee	0.7

Source: Dairy Facts, 2007, International Dairy Foods Association

Vanilla

Vanilla is without exception the most popular flavour for Ice Cream in North America. The dairy industry uses half of the total imported vanilla to North America. It is a very important ice cream ingredient, not only in vanilla ice cream, but in many other flavours where it is used as a flavour enhancer, e.g. chocolate much improved by presence of vanilla.

Vanilla comes from a plant belonging to the orchid family called Vanilla planifolia. There are several varieties of vanilla beans among which are Bourbon, Tahitian, Mexican. Bourbon beans are used to produce best vanilla extracts. Bourbons from Madagescar are the finest and account for over 60% of World production, Indonesia, 23%.

From each blossom of the vine that is successfully fertilized comes a pod which reaches 6-10 inches in length, picked at 6-9 months. It requires 26-29°C day and night throughout the season, and frequent rains with dry season near end for development of flavour.

Pods are immersed in hot water to "kill them" (also increases enzyme activity), then fermented for 3-6 months by repeated wrapping in straw to "sweat" and then uncovered to sun dry. 5-6 kg green pods produce 1 kg. cured pods. Beans then aged 1-2 yrs.

Enzymatic reactions produce many compounds-vanillin is the principal flavour compound. However, there is no free vanillin in the beans when they are harvested, it develops gradually during the curing period from glucosides, which break down during the fermentation and "sweating" of the beans. Extraction takes place as the beans are chopped (not ground) and placed in stainless steel percolator and warm alcohol (50°C, 50% solution) is pumped over and through the beans until all flavouring matter is extracted.

Concentrated Extract

Vacuum distillation takes place for a large part of the solvent. The desired concentration is specified as two fold, four fold, etc. Each multiple must be derived from an original 13.35 oz. beans.

Vanilla can be and is produced synthetically to a large extent. By-product of pulp and paper industry (lignan) or petrochemical industry (guaiacol). Compound flavours are produced from combination of vanilla extract and vanillin. Vanillin maybe added at one ounce to the fold and labelled Vanilla-Vanillin Flavour. Number of folds plus number oz. of vanillin equal total strength, eg. 2 fold + 2 oz. = 4 fold vanilla-vanillin. However, more than 1 oz to the fold is deamed imitation.

Vanilla flavouring is available in liquid form as:

- Natural Vanilla
- Natural and artificial (reinforced Vanilla with Vanillin)
- Artificial Vanilla (vanillin).

Usage level in the mix is a function of purity and concentration, usually ~0.3%.

Some vanillin actually improves flavour over pure vanilla extract but too much vanillin results in harsh flavours.

The choicest of ice creams can be made only with the best of flavouring materials. A good vanilla enhances the flavour of good dairy products in ice cream. It does not mask it.

Chocolate and Cocoa

The cacao bean is the fruit of the tree Theobroma cacao, (Cacao, food of the gods) which grows in tropical regions such as Mexico, Central America, South America, West Indies, African West Coast. The word cocoa is a corruption of the native word cacao. The beans are embedded in pods on the tree, 20-30 beans per pod. When ripe, the pods are cut from the trees, and after drying, the beans are removed from the pods and allowed to ferment, 10 days (microbiological and enzymatic fermentation). Beans then are washed, dried, sorted, graded and shipped.

At the processing plant, beans are roasted, seed coat removed-called the nib. The nib is ground, friction melts the fat and the nibs flow from the grinding as a liquid, known as chocolate liquor.

Liquor: 55% fat, 17% carbohydrate, 11% protein, 6% tannins and many other compounds (bitter chocolate-baking).

Cocoa butter: fat removed from chocolate liquor, narrow melting range 30 to 36° C.

Cocoa: after the cocoa butter is pressed from the chocolate liquor, the remaining press cake is now material for cocoa manufacture.

The amount of fat remaining determines the cocoa grade:

- medium fat (Breakfast) cocoa 20-24% fat
- low fat 10-12% fat.

Cocoa powder can also be alkalized, which reduces acidity/astringency and darkens the colour. Slightly alkalized cocoa is usually preferred in ice cream because it gives a deeper colour but the choice depends upon:

- consumer preference
- desired colour (Blackshire cocoa may be used to darken colour)
- strength of flavour
- fat content.

There are many types of chocolate that differ in the amounts of chocolate liquor, cocoa butter, sugar, milk, other ingredients, and vanilla.

Imitation Chocolate

Replacing some or all of the cocoa fat with other vegetable fats. Improved coating properties, resistance to melting.

White Chocolate

Cocoa butter, MSNF, sugar, no cocoa or liquor.

In chocolate ice cream manufacture, cocoa is more concentrated for flavouring than chocolate liquor (55% fat) because cocoa butter has relatively low flavour. However, the cocoa fat adds texture to the ice cream. Acceptable mixes can be made using 3% cocoa powder, 2.5% cocoa powder plus 1.5% chocolate liquor, or 5% chocolate liquor.

A good chocolate ice cream will be made if the cocoa and/or chocolate liquor is added to the vat and homogenized with the rest of the mix. Chocolate mixes have a tendency to become excessively viscous so stabilizer content and homogenizing pressure need to be adjusted.

One problem is called chocolate specking. It can occur in soft serve ice cream, when cocoa fibres become entrapped in the churned fat.

Fruit Ice Cream

Fruit for Ice Cream is available in the following forms:

1. Fresh Fruit
2. Raw Frozen Fruit
3. Open Kettle Processed Fruit
4. Aseptically Processed Fruit.

Advantages of processed fruits:

1. Purchasing year round supply: problems of procurement and storage transferred to fruit processor
2. Availability: blending of sources from around the world in RTU form, no thawing, straining, etc.
3. Quality control: processor adjusts for quality variations
4. Ice Cream quality: fruit won't freeze in ice cream, usually free of debris, straw, pits.
5. Microbial Safety
6. Convenience.

Fruit feeders are used with continuous freezers to add the fruit pieces, while any fruit juice is added directly to the mix. Fruit is usually added at about 15-25% by weight.

Nuts in Ice Cream

Nuts are usually added at about 10% by wt. Commonly used are walnuts, pecans, filberts, almonds and pistachios. Brazil nuts and cashews have been tried without much success.

Quality Control of Nutmeats for Ice Cream

1. Extraneous and Foreign Material: Requires extensive cleaning, Colour Sorter, Destoner, X-rays, Aerator, Hand-Picking, Screening
2. Microbiological Testing: Aflatoxin contamination can be a hazard with Peanuts, Pistachios, Brazils. All nutmeats should receive random testing for: Standard Plate Count, Coliform, E. Coli, Yeast and Mold, Salmonella.
3. Bacteria Control: Nuts must be processed in a clean sanitary premise following good manufacturing practices. Nuts should be either oil roasted or heat treated to reduce any bacteria.
4. Sizing: Some nutmeats require chopping to achieve a uniform size in order to fit through the fruit feeder, i.e.: Pecans, Almonds, Peanuts, Filberts

5. Storage Nutmeats should be stored at 34-38° F to maintain freshness and reduce problems with rancidity.

Colour in Ice Cream

Ice cream should have a delicate, attractive colour that suggests or is closely associated with its flavour. Almost all ice creams are slightly coloured to give them the shade of the natural product 15% fruit produces only a slight effect on colour. However, most suppliers, would include some colour in the fruit to save the processor time i.e. solid pack strawberries include colour. Most colours are of synthetic origin, must be approved, purchased in liquid or dry form. Solutions can easily become contaminated and therefore must be fresh.

Colours are used in ice cream to create appeal. If used to excess they indicate cheapness. The choice of shade is dictated by flavour, i.e. red for strawberry, light green for mint, purple for grape, etc.

Homemade Ice Cream

Ice cream fills a useful place in homes throughout the country. It is a favourite for desserts or snacks incorporating an array of many flavour variations. Ice cream contains many nutrients. With the recipes provided, all should be able to enjoy some type of this tempting food. If the regular recipe does not suit your needs, there is the low calorie recipe which contains less than 3% fat for both a cost and calorie saving.

The recipe using coffee whitener is significantly less costly than the regular and does not contain milk fat should that be your limitation. So let's mix up a batch of ice cream for anyone and everyone to enjoy!

Ingredients Used

The main constituents of ice cream are fat, milk solids-not-fat (skim-milk powder), sugar, gelatin (or other suitable stabilizer), egg and flavouring.

A variety of milk products can be used: cream, whole milk, condensed milk and instant skim-milk powder. The recipes stated below proved satisfactory using whipping cream (32-35% fat), table cream (18% fat) and whole milk. The fat gives the product richness, smoothness and flavour. Skim-milk powder is used to increase the solids content of the ice cream and give it more body. It is also an important source of protein which will improve the ice cream nutritionally. Use good quality, fresh powder to avoid imparting a stale flavour to the ice cream.

Liquid coffee whitener (usually purchased frozen) is a cream substitute in one of the recipes. It will yield a slightly different flavour which is still very acceptable. The texture of the ice cream is very creamy. Liquid coffee whitener offers the convenience of being stored frozen in your freezer and is readily available if a quick decision is made to make ice cream.

Sugar is a common ingredient to use as a sweetener. It increases the palatability and improves the body and texture. The next ingredient, gelatin (or similar substance) assists in absorbing some of the free water in the ice cream mix and helps prevent the formation of large crystals in the ice cream.

It also gives substance or a less watery taste when the ice cream is consumed. The eggs are added to make the fat and water more miscible and also to improve the whipping ability which gives the ice cream greater resistance to melting.

Preparation of the Ice Cream Mix

The mix (unfrozen ice cream) has to be cooked (pasteurized). For pasteurizing the mix, it is best to use a double boiler to prevent scorching.

Place the liquid ingredients (milk, cream or coffee whitener) in the upper section of the double boiler. Beat in the eggs and the skim-milk powder. Mix the gelatin with the sugar and add to the liquid with constant mixing. While stirring, heat to about 70°C. Place the container in cold water and cool as rapidly as possible to below 18°C.

Aging the Mix

The ice cream mix is best if it is aged (stored in the refrigerator) overnight. This improves the whipping qualities of the mix and the body and texture of the ice cream. If time does not permit overnight aging, let the mix stand in the refrigerator for at least four hours. After the aging process is completed, remove the mix from the refrigerator and stir in the flavouring.

Freezing the Mix

The freezing procedure has a two-fold purpose, the removal of heat from the mix and the incorporation of air into the mix. Heat is removed by conduction through the metal to the salt water brine surrounding the freezing can. This transfer of heat depends upon the temperature of the brine, the speed of the dasher and how well the dasher scrapes the cold mix from the surface of the freezer can. The dasher speed and surface contact are important to achieve complete removal of the frozen ice cream from the wall of the freezer can. A brine made from 500 grams (1 lb.) of salt and 5 kilograms (11 lbs.) of crushed ice (one pail full) makes a good freezing mixture.

Before starting to freeze the ice cream, make sure all parts of the freezer coming in contact with the ice cream are clean and have been scalded. Let the can cool before pouring in the mix. Place the empty can in the freezer bucket and insert the dasher ensuring both the can and the dasher are centred. Pour the cold, aged mix into the freezer can. The can should not be filled over two-thirds full to allow sufficient room for air incorporation.

The recipes listed below will fill a 5 litre (5 quart U.S.) freezer can to just below the fill line. Attach the motor or crank mechanism, depending on whether your freezer is the electric or hand-cranked style, and latch down securely. Plug in the motor or start turning the crank. Immediately begin adding crushed ice around the can sprinkling it generously with salt. Try to add the salt and ice in the same one to ten proportion to get the proper brine temperature. After the bucket is filled with ice to the overflow hole, pour a little water over the ice to aid in the melting process.

Freeze the mix for 20 to 30 minutes. If the electric motor stalls, immediately unplug it. Remove the motor or crank and take the dasher out of the ice cream. The ice cream will be softly frozen. Scrape the ice cream from the dasher and either scoop into suitable containers or pack in the freezer can. Immediately place the ice cream in the deep freeze to harden.

If freezer facilities are not available, the ice cream can be left in the can, the lid plugged with a cork and placed back into the bucket. Repack the freezer with more ice and salt, cover with a heavy towel and set in a cool place to harden until serving time. This will require further addition of ice and salt depending on the length of time the ice cream is being held. The yield from the recipes listed below should be three to four litres.

Regular Vanilla Ice Cream

Table cream 2 litres (2 US quarts)

Instant skim-milk powder 350 ml (1.5 cups)

Sugar 450 ml (2 cups)

Gelatin one 7 g (1/4 oz.) pkg.

Egg one med or large

Vanilla 10 ml (2 teaspons)

Calories per 100 g 230

Low Calorie Vanilla Ice Cream

Whole milk 2 litres (2 US quarts)

Instant skim-milk powder 500 ml (2 cups)

Sugar 350 ml (1.5 cups)

Gelatin one 7 g (1/4 oz.) pkg.

Egg one med or large

Vanilla 10 ml (2 teaspons)

Calories per 100 g 125

Milk Substitute Vanilla Ice Cream

Coffee whitener 2 litres (2 US quarts)

Instant skim-milk powder 350 ml (1.5 cups)

Sugar 500 ml (2 cups)

Gelatin one 7 g (1/4 oz.) pkg.

Egg one med or large

Vanilla 10 ml (2 teaspons)

Calories per 100 g 210

Hints for Making Good Ice Cream

1. If the ice cream is very soft, the brine is not cold enough. More salt should be added to reduce the brine temperature.
2. If the ice cream is coarse and ice in less than 20 minutes, the brine has become to too cold too quickly. Too much salt has been used.
3. Make the ice cream mix the day before it is frozen to get a smoother product and a higher yield.
4. Electric freezing takes longer than hand operated.
5. Use crushed ice for freezing.
6. Freeze at least 3 hours before the ice cream is to be served.
7. Be sure dasher is properly centred in the freezer can.
8. Add liquid flavours before freezing but if you want to add fruits or nuts, add them after freezing and before hardening.
9. Use a wire whip to blend ingredients for best results.
10. Clean the salt off all the metal parts of the freezer to prevent corrosion.

Milk and Cheese

In food preparation, milk is used in many ways and combined with many kinds of foods. All the different types of cheese are made from it.

Meat, vegetables, and cereals may be cooked in it. It is used as a basis for many sauces. Such sauces may be combined with eggs, with meats, or with vegetables; it is used in puddings and in frozen desserts; it is used for soups and for drinks like cocoa and coffee; it is combined with cereals; it is used in combination with many foods, as in custards, in cakes, and in quick breads.

Milk from different animals is used for food, but in this country unless the source of the milk is mentioned it is understood to be cow's milk.

Homogenization of Milk

Globules of butter fat are suspended in the milk. They are surrounded by films of adsorbed caseinates, albuminates, and globulinates. The fat globules of milk are too large to form a permanent emulsion, so they gradually rise to the top of the milk in the form of cream. If the milk or cream is put through a machine called a homogenizer, the fat globules are reduced in size. This is accomplished by using pressure and forcing the milk or cream through small openings. Homogenized milk or cream may form a stable emulsion if the fat globules are reduced enough in size. Hence, when the fat is broken into fine enough globules the cream will not rise to the top of the homogenized milk.

The size of the fat globules after homogenization depends upon the temperature of the milk during homogenization and the pressure used. With increase in temperature the degree of dispersion increases rapidly from 40° to 65°C, so that the smallest fat particles are obtained at 65°. Ordinarily temperatures above 65° are not used for homogenization. The size of the fat particles also decreases with increased pressure.

Whipped cream is stabilized by proteins. The fat globules in cream are surrounded by films of protein substances. Homogenized cream also has the film of adsorbed proteins around the fat particles. Clayton states the fat particles in homogenized cream may be 1000 times greater in number than before homogenization. Since the number of fat particles is increased, the amount of globulinates, caseinates, and albuminates used in forming films is very much greater, for the surface area of the fat globules has increased enormously.

Whipped cream is both an emulsion and a foam. The fat particles must be surrounded by a film of protein in order to be stabilized, and the air globules must be surrounded by a film of protein to stabilize them. In homogenized cream most of the protein is used in surrounding the fat globules, on account of the increased surface area of the smaller and increased number of fat globules, and thus there is not enough left to surround the air bubbles. Hence, homogenized cream seldom whips unless protein is added for film forming.

Factors that affect the whipping quality of cream. In addition to the protein or film forming in whipping cream, the fat content, the size of the fat particles, the temperature of whipping, and the viscosity are important factors. Dahlberg and Hening have studied the relation of viscosity, surface tension, and whipping properties of milk and cream. They have found that increased viscosity increases the whipping properties of cream, but the lowering of the surface tension does not improve the whipping qualities. They have reported two changes taking place during whipping. The incorporation of air depends upon the milk proteins forming the film around the air globules, and the rigidity or stiffness of the whipped cream depends upon the clumping together of the fat particles. The best whipping cream did not give as large a volume as some other creams, but it had less liquid drain out of it after whipping.

Cream whips better with an increasing fat content up to 35 per cent. The cream with the higher fat content gives more particles for clumping and also increases the viscosity of the cream.

As the fat particles clump at the liquid/air interface, or within the liquid, the increased rigidity they give the foam permits inclusion of more air bubbles and extension of the films with the result that the dryness of the foam is increased. Larger fat particles clump more readily and thus form the structural support offered by the fat more easily. This offers an explanation of why cream from milk containing larger fat particles, milk from Jersey and Guernsey breeds, whips more quickly than cream containing smaller fat particles, milk from other breeds.

Aging improves the whipping qualities of cream. The viscosity increases with aging. As a general rule, treatment that increases the viscosity increases the whipping properties. Pasteurization tends to reduce the whipping quality of cream.

Babcock has found that the best whipping is obtained at a temperature of 45°F. or lower. At this temperature agitation favours the clumping together of the fat particles. At high temperatures, both the higher temperature and the agitation increase the dispersion of the fat. Above 50°F. the decrease in stiffness of whipped cream is in direct ratio to increase in temperature, so that 30-per cent cream will not whip at 72°F.

Babcock found acidity up to 0.3 per cent, at which sour taste is evident, had no effect on whipping quality. If acid was added in excess of 0.3 per cent, whipping quality improved, whether added to fresh or aged cream, when the amount added began to curdle the cream.

The addition of sugar to cream either before or after whipping was found by Babcock to decrease the stiffness of the cream. For each 2 tea-spoons added to 100 cc. (about 1/2 cup) the stiffness decreased four points on the stiffness scale. Adding the sugar before whipping the cream decreased the volume obtained and increased the whipping time.

The denaturation or coagulation of the protein film at the air/cream interface, increases the stiffness of the whipped cream. Since colloidal reactions require time, it is a better practise to add sugar to whipped cream after, rather than before or during the whipping process. By this procedure denaturation is more complete, which offers an explanation of why the volume of the cream is less affected by the addition of the sugar last. The addition of sugar lessens the stiffness and decreases the volume of the whipped cream because it either prevents denaturation and/or peptizes the protein film.

Reaction of Milk

Freshly secreted milk is nearly neutral to litmus. The reaction varies slightly but has an approximate pH of 6.6. The freshly secreted milk contains carbon dioxide. The amount of this gas in the milk decreases during milking and the subsequent handling of the milk, while the percentage of oxygen and nitrogen increase. For this reason the titratable acidity decreases for a time in milk exposed to the air. Confined milk does not show as great a decrease in titratable acidity as the exposed milk, for the percentage of carbon dioxide lost is smaller.

Effect of heating milk on acidity. When milk is heated at the boiling point or at temperatures above or near the boiling point the titratable acidity at first decreases owing to the loss of carbon dioxide, and then increases. Whittier and Benton report that the hydrogen-ion concentration increases continuously. They find the hydrogen-ion increase and the later increase in titratable acidity is due to the formation of acids from constituents of the milk. The amount of acid produced depends upon the time and temperature of heating, a greater amount of acid being produced with a longer heating period and with higher temperatures. From their experiments they conclude that the acid is produced from the lactose of the milk. They have shown that, the greater the concentration of lactose present, the greater the amount of acid formed at a definite temperature and for a definite time.

Coagulation of Milk

Under certain conditions, the addition of alcohol as well as the application of heat may cause coagulation of milk. Milk may be coagulated by the addition of rennin or by bringing the acidity of the milk to the isoelectric point of the casein. When milk is combined with other foods, the salt content of the food or the tannin content of the food may be factors that aid coagulation.

Alcohol Coagulation: The addition of 70 per cent alcohol to milk may cause coagulation. As the pH of the milk decreases, it becomes susceptible to alcohol precipitation, though this varies with different milks. Usually the milk is precipitated by alcohol while it is still stable to heat and sterilizing temperatures. Some freshly secreted milk is coagulated by alcohol, but this milk is usually abnormal in some way. About 2 cc. of alcohol are added to an equal quantity of milk for the test. Casein is precipitated by alcohol as calcium caseinate, calcium is not released as by acid coagulation.

Rennet Coagulation: Milk may be coagulated by the addition of rennet. Rennet is an extract that is usually obtained from the inner lining of the stomachs of calves and lambs. The rennet contains an enzyme called rennase or rennin. The clotting of the milk is generally believed to be the direct action of the rennin on the casein. But the manner in which these changes is produced is not fully understood. A very small amount of rennin is capable of coagulating a large amount of milk. At favourable or optimum hydrogen-ion concentrations for clotting, 1 part of the fairly pure enzyme preparation is able to coagulate 3,000,000 or more parts of milk.

Mechanism of Clotting: It is usually stated that the casein is changed to paracasein by the action of the rennin. It is also often stated that the clotting is brought about in two steps, the first being the action of rennin on the casein and the second the precipitation of the changed casein. Rogers reviews the many theories of rennin coagulation. Some investigators claim the changes are purely chemical; others maintain the rennin affects only the physical state of the calcium caseinate. However, if the change can be explained on the basis of colloid chemistry, it is probable that absorption and the electric charge play an important role in the process. Rogers states that Hammarsten regards casein in milk as a calcium caseinate-calcium phosphate complex. "As a matter of fact the compound called calcium caseinate is most probably a true calcium phosphocaseinate, if, as seems likely, the second and third hydrogens of the orthophosphoric acid esterfied with certain of the amino acids in the casein molecule react with calcium. The correct conception of the term 'calcium phosphocaseinate,' as it is now commonly employed, is that of a colloidal calcium phosphate (or phosphates) sol protected by a calcium caseinate (or caseinates) sol in a manner as yet imperfectly understood." The stabilization of sols is best explained by the theory of Helmholtz, i.e., each colloidal particle is surrounded by an electrical double layer. "In the case of negatively charged sols, in which class calcium caseinate and calcium paracaseinate evidently fall, the outer layer consists of hydrogen ions. If these are replaced by a sufficient number of positively charged ions of higher charge, e.g., calcium ions carrying two positive charges"; or, in other words, if these ions are more strongly adsorbed than the hydrogen ions, the colloid particle will readily precipitate, the rate of clotting being determined by the rate of replacement.

Richardson and Palmer state that rennin itself may reduce the charge of the calcium caseinate micelle and thus reduce the stability of the casein sol. They found indications that the isoelectric point of rennin is about pH 6.9 to 7.0. Above this pH the rennin is negatively charged and below pH 6.9 it is positively charged. They found that rennin lowered the electro-phoretic velocity of calcium caseinate and calcium phosphocaseinate micelles when the casein sol was negatively charged and the rennin was positively charged, but not when the rennin was negatively charged (above its isoelectric point, pH 6.9 to 7.0). From this evidence and from the fact that paracaseinate micelles are not affected by rennin, which agrees with the fact that casein once coagulated by rennin has lost its sensitiveness to this enzyme, they suggest that rennin acts by sensitizing the casein by a preliminary reduction of the electric charge on the casein micelles.

During the clotting of the milk, aside from the consistency of the milk, there is little change in its physical properties. The hydrogen-ion concentration does not change during the clotting process.

Factors affecting action of rennin. Several factors influence the activity of the rennin in bringing about coagulation. These may be listed as follows: (1) temperature for rennin action; (2) heating the milk before the addition of rennin; (3) hydrogen-ion concentration; (4) concentration of casein, calcium, and phosphate ion; (5) character of cations used for coagulation.

Temperature for rennin action. The optimum coagulation by calf rennin is about 40° to 42°C. Below this temperature coagulation is less rapid and no clotting occurs below 10° to 15°C. Also no clotting occurs above 60° to 65°C. The clot is softer at low temperatures and tougher and stringy at high temperatures. By optimum is meant the temperature at which coagulation takes place most rapidly for a definite concentration of rennin and milk. Effect of previously boiling the milk upon rennin coagulation. If milk is boiled and then cooled before the rennin is added, the rate of coagulation is retarded and a much softer, more flocculent clot is obtained. Pasteurization also affects the rate of coagulation of the milk and the type of clot formed by rennin but not to the extent that boiling does.

Richardson and Palmer found by electrokinetic evidence that heat increased the electric charge on the casein micelles or the cataphoretic velocity of the casein solution. The fact that rennin does not form as firm a clot with milk that has been previously heated indicates that rennin reduces the charge on the casein particles but not sufficiently to form a firm clot. This offers a colloidal explanation of why the addition of active cations (as calcium chloride) to heated milk causes the rennin to coagulate the milk normally.

Hydrogen-ion Concentration: The reaction of the milk affects the rapidity of coagulation and the character of the curd formed. Ordinarily when the reaction of the milk is alkaline coagulation does not occur. This is shown by the addition of a small amount of soda to milk before the addition of junket. The optimum hydrogen-ion concentration for rennin activity has been reported to lie in the zone between pH 5.99 and 6.40.

Character of Cations: In addition to rennin, cations are necessary to bring about coagulation of milk. Because casein and calcium are so closely involved in milk, the cation calcium is important in bringing about coagulation. Hence, Rogers states that it is to be expected that the concentration of both casein and calcium markedly affect both the rate of coagulation and the character of the clot. If milk is diluted with sufficient water, clotting is both delayed and incomplete, the clot being soft. If calcium chloride is added to the water, diluted milk clotting properties are restored, which suggests that the concentration of calcium ions is more important than that of the casein ions.

Rogers states that any metallic ion can replace the calcium in coagulation. However, it is generally accepted that the sodium and potassium salts of paracasein are soluble. Monovalent ions are less effective than divalent ones in replacing the calcium. Rogers reports that all monovalent ions did not bring about coagulation in some instances. The divalent ions were not all equally effective, calcium and barium being more efficient than magnesium.

Sugar: Sugar tends to prevent the coagulation of milk by rennin.

Coagulation of milk by acid. Kruyt states that there are some proteins that are not sufficiently hydrated to be stable by hydration alone. He cites casein as an example of a protein "which can exist either in acid or in an alkaline solution, but does not dissolve in water, with the consequence that the sol ordinarily flocculates when neutralized." Either the acid produced during fermentation or acids added to milk precipitate the casein. The casein is least soluble at its isoelectric point pH 4.6. If enough acid is added to lower the pH below 4.6, casein salts, such as casein chloride or casein lactate, are formed. If these salts are soluble, the casein goes into solution. Hence the largest yield of precipitated casein is near the isoelectric point.

Fermentation of Milk: Fermentation, or the production of lactic acid from lactose by bacteria, takes place in milk that is allowed to stand under favourable conditions. Rogers states that true lactic acid fermentation is brought about by the Streptococcus lactis and certain other organisms, lactic acid being the principal end-product, other products being present in only small amounts. In mixed lactic acid fermentation, or when other organisms in addition to S. lactis are present, the end-products may include acetic, propionic, lactic, succinic, formic, and butyric acids, carbon dioxide, hydrogen, acetone, and ethanol. As fermentation increases, an acidity is reached at which the action of most bacteria is suppressed. When fermentation is checked at pH 4.8 to 5.0, the bacteria consists chiefly of Streptococcus lactis.

The rate of fermentation depends chiefly upon the temperature at which the milk is held. At low temperatures, on account of retardation of bacterial action, it takes place slowly. Rogers states that fermented milk, allowed to stand at a fairly high temperature, undergoes a second lactic acid fermentation brought about by the Lactobacillus bulgaricus organisms. Some of these types of bacteria form a high percentage of acid and the hydrogen-ion concentration may reach pH 3.23.

Changes occurring during acid precipitation. During fermentation chemical and physical changes occur in the milk. The flavour becomes acid. The calcium caseinate is changed to casein. During this process calcium is split off and forms soluble calcium lactate. In addition some dicalcium phosphate is converted into monocalcium phosphate. Curdling or clotting occurs when the acidity reaches about pH 5.3. During the clotting process the hydrogen-ion concentration does not increase. Milk clotted by fermentation is often called clabbered milk. Its flavour and aroma may vary, depending upon the types of bacteria producing the fermentation. Fermented milk may be used for drinking, for cooking, and for cottage cheese.

Cheese, such as cottage cheese, when clotted by acid coagulation, loses a large proportion of its calcium. The calcium salts become soluble more rapidly than the phosphorus; hence a larger proportion of the calcium than of the phosphorus is lost in the whey. Casein precipitated by rennin retains most of its insoluble salts, hence has a larger proportion of calcium than the acid precipitated casein.

Heat Coagulation: The term heat coagulation refers to the so-called "denaturation" of the protein, by which it is rendered insoluble.

Lactalbumin. Lactalbumin has temperatures for heat coagulation similar to that of egg albumin. The lactalbumin forms a flocculent precipitate, whereas egg albumin forms a firm coagulum. Rupp has reported the following amount of lactalbumin coagulated when heated for 30 minutes.

Temperature, °c.	*Albumin rendered insoluble, per cent*
62.8	0.00
65.6	5.75
68.3	12.75
71.1	30.78

Casein. Casein is not coagulated by heat at ordinary temperatures or when heated for short periods, though the heating may alter the casein. Rogers states that it is necessary to heat milk about 12 hours at 100°C. to bring about coagulation. It takes approximately 1 hour at 135°C. and approximately 3 minutes at 155°C. The time and temperature vary somewhat with different milks.

The rate of coagulation depends upon the concentration of the casein as well as the time and the temperature of heating. Rogers states that evaporated milk, containing twice the concentration of solids-not-fat in normal milk, and thus a higher concentration of casein, requires about 60 minutes for coagulation at 114.5°C, 10 minutes at 131 °C, and 7500 minutes at 80°C.

Rogers and Palmer both state that, in the evaporated-milk industry, the forewarming of milk prior to processing increases its stability to heat. "Rapid improvement in resistance to heat coagulation results in increase in temperature for prewarming up to 90° to 100°C. Above 90°C. the change is very small, but in some cases can be effected with increases in temperature up to 120°C. for 10-minute periods of forewarming. When time is chosen as the variable, improvement may be noted with increases in the time up to 30 minutes at a temperature of 95°C. At higher temperatures the same improvement may be effected in shorter periods of time."

Fat: Rogers states that fat particles in relatively large aggregates may act as nuclei about which coagulation of the casein can proceed. In un-homogenized milk the fat affects the coagulation time and temperature but slightly. But when a milk of higher fat content is homogenized the fat clumps may act as nuclei about which the casein may gather during heating. With increase in homogenization pressure as well as fat content, other conditions being the same, a marked decrease in stability to heat is noted. In homogenized milk it was found that the maximum stability to heat coagulation occurs if homogenization is carried out at 80°. Rogers adds that the feathering of some homogenized cream when added to coffee may be caused by using too high homogenization pressure, thus reducing the stability of the cream to heat.

The role of salts in heat coagulation of milk. In heat coagulation of milk, the milk salts play an important role, for the salt equilibrium is altered by heat. When milk is boiled precipitation of part of the calcium phosphate occurs. Sommer and Hart have concluded that salts are the main factor in heat coagulation of fresh milk. Electrolytes have a marked effect upon the stability of colloids. In precipitating a hydrophilic colloid divalent and trivalent ions are generally more effective than monovalent ones. In the milk are found the monovalent cations, sodium and potassium; the monovalent anion, chlorine; the divalent cations, calcium and magnesium; and the trivalent anions, phosphate and citrate. Sommer and Hart concluded that the coagulation of milk on heating may be due to an excess or a deficiency of calcium and magnesium.

They explain this as follows. "The casein of the milk is most stable with regard to heat when it is in combination with the calcium. If the calcium combined with the casein is above or below this optimum, the casein is not in its most stable condition. The calcium of the milk distributes itself between the casein, citrates, and phosphates chiefly.

If the milk is high in citrate and phosphate content, more calcium is necessary in order that the casein may retain its optimum calcium content after competing with the citrates and phosphates. If the milk is high in calcium there may not be sufficient citrates and phosphates to compete with the casein to lower its calcium content to the optimum. In such cases the addition of citrates or phosphates makes the casein more stable by reducing its calcium content. The magnesium functions by replacing the calcium in the citrates and phosphates."

Heat coagulation of casein endothermic. Leighton and Mudge have shown that an endothermic reaction accompanies the appearance of visible curds when milk is coagulated by heat. This is accompanied by precipitation of calcium and magnesium as phosphate and citrates. A similar reaction occurs in custard.

In cooking custard, the ingredients of which are milk, egg, and sugar, the temperature drops or does not rise for a period of time during coagulation or setting of the custard, a condition particularly noticeable just before curdling takes place.

Coagulation of milk by cooking meat or vegetables in it. Fresh milk is seldom coagulated by heating for home use. The temperature attained in ordinary heating is not great enough to cause coagulation, nor is the milk heated for the long period required for coagulation at boiling temperatures. But with the addition of other foods to milk in food preparation, coagulation often occurs with a very short period of heating. One of the factors in this coagulation is undoubtedly the salt content of the food added to the milk as well as the salt content of the milk.

The balance of the milk salts for greatest stability may be upset and coagulation occurs when the food is heated in the milk. Cooking of meat in milk. Ham is often baked in milk. Sometimes pork chops are floured, seared in fat, and then baked in milk. Other meats and fish are sometimes baked or cooked in milk. Often curdling of the milk occurs, and the appearance of the meat, owing to adherence of curds of milk, is not attractive. Thus it is desirable to prevent curdling. Among the causes of curdling are the temperature at which the meat is cooked, the salt content of the food cooked in the milk, the manner in which the milk is added to the meat, and the reaction of the milk.

The higher the temperature at which the meat is cooked the greater the tendency to curdle. Larger curds are also formed at higher temperatures. The temperature to which milk must be heated to bring about curdling is high, so this cooking temperature alone is not sufficient to bring about the coagulation. Even when the meat is cooked in an oven at a high temperature, the liquid portion does not reach a higher temperature than boiling. It is possible that the acidity is increased during cooking, but the resulting pH of the meat-milk broth is changed very little from that of the original milk or from that of the meat broth when the meat is cooked in water. Thus it seems that the heating and the acidity developed during cooking do not alone bring about coagulation. The salt content of the food cooked in the milk probably influences the coagulation, and this combined with the heating, the temperature of heating, the acidity developed, and the altering of the casein by heat are sufficient to cause curdling. Sodium chloride has some effect, for curdling is more likely to occur when the meat cooked in the milk is a cured or salted one than when fresh meat is used.

Curdling may be prevented by the addition of soda, about 1/16 teaspoon per cup of milk. From this it appears that the reaction has some part in the coagulation. The soda may combine with other salts that tend to bring about coagulation or the coagulation may be prevented by the slightly alkaline reaction. A slightly alkaline reaction also prevents coagulation by rennin and by fermentation.

If a portion of the milk is added to the meat when cooking is first started and the rest of the cold milk added gradually to the meat during the cooking period, curdling is less likely to occur. The addition of acid foods, such as prepared mustard, which may contain vinegar, or apples and pears to be baked with the meat, would tend to increase the tendency to curdle. Evaporated milk has less tendency to curdle than fresh milk, which may be due to the previous heating.

Cooking of vegetables in milk. Milk usually does not curdle when cabbage, chard, spinach, or cauliflower is cooked in it. But it is likely to curdle when asparagus, string beans, peas, and carrots are cooked in it. Asparagus usually curdles the milk after a few minutes of cooking. There are several factors that may aid in bringing about coagulation of the milk. The slight acidity of some vegetables combined with the heating of the milk may tend to bring about coagulation, but the acidity is not usually great enough, nor the boiling temperature high enough, nor the boiling long enough continued for these factors to be very important. The salt and the tannin contents of the vegetables are probably the principal causes of coagulation.

Some vegetables contain larger amounts of tannin than others. Tannin is a dehydrating agent and brings about denaturation of hydrophilic sols, like gelatin, starch, and agar-agar. After denaturation the hydrophile is sensitive to small amounts of electrolytes and precipitation occurs readily. Kruyt states that tannins do not bring about dehydration of the protein in an alkaline medium.

Hence the addition of soda in small amounts to the milk in which the vegetable is cooked prevents coagulation of the milk. Tannins lower the surface tension, which results in foaming of vegetables containing tannin when they are cooked in water. It is rather interesting that the vegetables that usually foam the most when cooked in water are the ones that have the greatest tendency to coagulate milk.

Tomato soup. When tomatoes are combined with milk to make cream of tomato soup, coagulation may occur. The acidity of tomatoes varies somewhat, but is about pH 4.4 to 4.6. If the amount of tomato added to the milk is great enough to lower the pH of the mixed milk and tomato to 4.8 to 4.6, the casein is precipitated without heating. This may happen if the milk is already fairly acid.

Since a longer time of heating milk increases the tendency to curdle it is preferable to heat tomato soup for only a short time. Heating slowly also increases the tendency to curdle, which may be due to the longer time required. In Experiment 55 in the laboratory outline several different methods of combining the tomato juice and milk are given. The tomato is usually added to the milk by stirring, for in this way the milk is diluted with a smaller amount of acid substance during the first part of the mixing. There is less tendency to curdle when the hot tomato is added to cold milk, than when the cold tomato is added to hot milk. Probably the slight denaturation brought about by heating the milk may partially account for this. Occasionally some milk is acid enough and the tomato is acid enough to cause curdling with all methods of combining unless soda is added. At other times curdling does not occur with any method of combining as outlined in this experiment.

Fruits and milk. When cream is added to fruit, clotting often occurs. This is usually due to the acidity of the fruit, but may also be due to an enzyme in it. Raw pineapple contains an enzyme, bromelin, that brings about clotting of milk. However, the pineapple juice not only brings about clotting but also peptization, for after a time the clot formed is less firm and the flavour is similar to that of peptized meat.

Boiling and Heating of Milk

The physical and chemical properties of the constituents of milk account for the behaviour of milk during its use in food preparation. Thus substances that lower surface tension become concentrated in the liquid/air interface. Proteins lower the surface tension of aqueous sols, hence accumulate in the surface. When milk is heated in an open pan, a scum or skin forms over the surface of the milk. At first this skin is rather thin and mobile but is gradually altered so that it becomes tenacious and tough enough to be removed with a stirring rod or spoon. This scum has been said to contain coagulated albumin and globulin. Tinkler and Masters state that if the scum is removed as formed, the total amount of protein that can be removed exceeds the total amount of albumin and globulin in the milk. When foods are cooked in milk the milk not only foams readily but the scum tends to hold the steam formed in heating the milk; it is because of this that the milk "boils over" so readily.

Sugar Reactions with Proteins of Milk

Ramsay, Tracy, and Ruehe investigated the substitution of dextrose for sucrose in sweetened condensed skim milk. They found the objections to using dextrose were (1) a brown discoloration, (2) a physical thickening, and (3) crystallization of the dextrose during storage. The last objection could be remedied by using 50 per cent dextrose and 50 per cent sucrose. The progressive thickening during storage at high temperature was caused by action of the dextrose on the casein and albumin of the milk.

During this investigation they found additional evidence that sugars react with proteins. When dextrose, lactose, or levulose was heated with skim milk or freshly precipitated casein, a dark brown colour formed in the product. When the sugars were heated in distilled water solutions to 250°F. for 30 minutes no caramelization occurred. Neither did darkening occur when albumin or casein was heated in water solution. But when the milk, albumin, or casein was heated with lactose or dextrose, a brown discoloration occurred. As the temperature was raised the dextrose and casein became so firmly attached to each other that no amount of washing could remove the sugar. Most of the biruet action of the skim milk was lost. The results were explained on the basis that a protein-sugar complex of glucosidal nature was formed.

On heating amino acids with dextrose highly coloured products were formed, the reaction probably being a condensation of an amino acid with an aldehyde or ketone group of the sugar. The very stable linkage of the aldehyde group of the dextrose and the ketone group of the fructose in the sucrose molecule is cited to explain the failure of sucrose to form condensation products with casein, albumin, or amino acids. It was found that as the reaction became more alkaline the appearance of the brown colour was more rapid.

The increased alkalinity was said to favour the change of the sugar from a lactone to a free aldehyde form, the free aldehyde acting with the amino acid or-NH1 groups of the protein. If the pH was much above 7 the milk was almost black. Hence, the sugar used, the reaction, and temperature all influenced the development of the brown colour. The length of heating in connection with the temperature was important as relatively high temperatures for a short period gave only slight development of the brown colour. It is almost impossible to retain the natural colour of fresh milk in the condensed milk products, for some brown discoloration occurs in the unsweetened and sweetened product whether made from whole or skim milk.

Whittier and Benton had shown that the hydrogen-ion concentration increases at a rate which is the function of the lactose concentration and the time and temperature of heating. Or, in other words, when milk is heated for a sufficient time at high enough temperatures the lactose is decomposed with formation of acid products. Hence, when milk is heated with sucrose the increasing acidity inverts some of the sucrose to dextrose and levulose, with the development of a brownish colour. One example of this is in the cooking of caramels, more brown colour developing with long slow cooking of the sucrose and milk. Another instance where this is used to advantage is in making caramel pudding by boiling, in the can, sweetened condensed milk for three hours or longer. The can and contents are chilled. On opening the can it is found that the contents have developed the brown colour of caramelized products and are thickened to the consistency of a pudding. This combination of sugar with milk proteins to form a thickened product is interesting in view of the fact that sucrose, dextrose, and levu-lose prevent the heat coagulation of egg albumin.

The housewife also makes use of the effect of acid on sweetened condensed milk. If about 1/2 cup of lemon juice is stirred into the contents of a can (about 1 1/2 cups) of sweetened condensed milk, the mixture thickens to a consistency that can be used for a pudding or pie filling and may be thinned with water to a desired consistency. The explanation of the thickening lies in the action of the acid on the complex sugar-protein combination.

Cheese

Definition: The Food and Drug Administration defines cheese, in the regulatory announcements, as "a product made from curd obtained from the whole, partly skimmed, or skimmed milk of cows, or from milk of other animals with or without added cream, by coagulating with rennet, lactic acid, or other suitable enzyme or acid, and with or without further treatment of the separated curd by heat or pressure, or by means of ripening ferments, special molds, or seasoning."

Classification of cheese. Cheese may be classified in many ways as (1) method by which the curd is produced, i.e., acid or rennet coagulation, (2) source of the milk, from cow, sheep, or goat, and (3) the texture and consistency of the cheese, i.e., whether soft, semi-hard, or hard. Many other classifications might be used but none of them are entirely satisfactory. Doane and Lawson list and describe nearly 300 cheeses. They state there are probably about 18 distinct varieties of cheese.

For purposes of discussion, Rogers classifies cheese as follows:

Soft

- Unripened Cottage Cream Neufchatel Ripened
- Ripened by molds
- Camembert
- Brie Ripened by bacteria
- Limberger
- Liderkranz.

Hard

- Semi-hard
- Ripened by molds
- Gorgonzola
- Roquefort
- Stilton Ripened by bacteria
- Brick
- Munster Very hard
- Without gas holes
- Cheddar
- Edam
- Gouda With gas holes
- Emmenthal
- Swiss
- Parmesan.

Composition of cheese. From the standpoint of quantity the principal constituents of cheese are casein, fat, and water. In addition it contains various salts, and unless heated to pasteurization temperatures various organisms such as bacteria and molds. Different types of soft cheese may contain from 40 to 75 per cent of water, hence this type does not keep long. Hard-type cheeses usually average 30 to 40 per cent moisture. The soft types may contain from 13 to 21 per cent of protein and from 0.5 to 50 per cent of fat. Hard types contain from 20 to 45 per cent protein and 19 to 40 per cent fat.

Coagulation of Milk for Cheese: Coagulation may be brought about by rennet or acid. Rennet-formed curds are more elastic, the acid ones more sticky. In acid-formed curds more of the calcium salts are split off from casein, forming calcium chloride which is soluble in the whey. Rennet-coagulated cheeses of cheddar types retain about 80 per cent of the calcium of milk, whereas soft cheeses retain about 20 per cent.

The temperature for coagulation varies with the type of cheese desired. In general, the lower the temperature the softer the curd. Curds formed at 21° to 25°C. are used for some soft cheeses. Cheddar cheese has a firmer curd and the milk is brought to 30°C. before the starter and rennet are added. Temperatures as high as 48°C. may be used for some cheese, the curd produced being distinctly tough and somewhat rubbery and elastic.

Making Cheese: The essential steps in making cheese are: blending the particular type of milk desired; bringing the milk to a definite temperature; adding lactic acid culture for types of cheese that need greater hydrogen-ion concentration when the rennet is added (acid cultures are added to cheddar types, but not to Swiss); adding vegetable colour, if cheese is to be yellow, omitting if cheese is to be American white; and adding the rennet. After coagulation the curds are cut to the definite size for the type of cheese desired.

Small curds retain less moisture within the curd but the whey does not drain so well from the curd. The next step is stirring the curd gently to facilitate draining of the whey. The curd is then ditched, salted, put in molds lined with cloth, and pressed into definite shapes as Longhorms, Prints, Daisies, Flats, Twins, and Cheddars. After being pressed the cheese may be soaked in salt brine or dry salt may be rubbed on the surface. Sometimes no additional salting occurs.

Soft unripened cheese is not cured; but after being pressed or moulded other types are placed on shelves in caves or specially constructed curing rooms to ripen. In the latter ventilation, humidity, and temperature may be carefully controlled according to the type of cheese. The curing period varies for different types of cheese and for the same type. For example, Cheddar may be cured from 2 or 3 months to 2 years. With longer curing a sharper, richer, and fuller flavour is developed.

Cheese, after being cured, is often blended for uniform flavour, texture, and body.

Secondary Heating of the Curd: A secondary heating of the curd is necessary with most hard and semi-hard cheeses. Making Emmenthal involves heating to about 55° to 58°C. This heating hastens the driving of the whey from the curd, changes its texture, and often alters the bacterial flora. The heating at high temperatures decreases the moisture content and rennet action is checked if not wholly stopped. Various physical changes take place during this period, the curd becoming tough, firmer, and rubbery. In Swiss and Parmesan cheese it also acquires plasticity.

Ripening of Cheese: In the process of ripening chemical and physical changes occur in the cheese. It loses its tough, rubbery qualities and becomes soft and mellow, sometimes almost crumbly. During this change as much as 50 per cent of the nitrogenous constituents may be converted to soluble forms, though the average for hard cheese is 30 per cent. These changes not only alter the texture and flavour, but also alter the cooking quality of the cheese, the increased solubility of the proteins increasing the ease with which the cheese may be blended with eggs, milk, and white sauce.

Ripening is slower at lower temperatures and more rapid at higher ones. Not only enzymes of the milk, if the milk has not been heated to a temperature to destroy the enzymes, but bacteria aid in ripening of the cheese and hydrolysis of the proteins. Some bacteria, such as lactic acid, produce enzymes that split the protein. More hydrolysis occurs in the softer centre of hard cheese than near the rind. Salting affects the rate of ripening by delaying bacterial growth, the proteins of cheese with more salt becoming soluble at a slower rate. Salt penetrates slowly from the rind to the centre and aids in drying the cheese. Changes in the fat in the interior of most cheese are usually negligible.

For the growth of molds and aerobic bacteria, holes must be punched in the cheese to allow oxygen from the air to penetrate. In the early stages of ripening Emmenthal and Swiss cheeses are soft and become elastic. It is during this stage that the holes or "eyes" are formed from production of gas, principally carbon dioxide, if ripening is normal, but with more hydrogen in abnormal or early ripening. If the cheese becomes too firm before the formation of holes is complete, checks and cracks appear in the cheese.

Cheddar Cheese in Cans: The Bureau of Animal Industry (Rogers) has announced a practical method of canning unripened Cheddar cheese. By this method a one-way or check valve, which holds perfectly against external pressure but with internal pressure allows gases formed during ripening of the cheese to escape, is inserted in the lid of the can.

Cheddar cheese has always been pressed in cylindrical forms of varying sizes, but in general rather large. When these large cheeses are cut they lose moisture, so the cut surface dries rapidly. In addition, if the cheese is well ripened, loss occurs through crumbling. In packing cheese in cans, the cheese, after pressing, is cut into the desired shape. Since hydrogen sulfide is often liberated during ripening of cheese, it is preferable to wrap the cheese in parchment and it is necessary to use a lacquered can, for the hydrogen sulfide tends to form a black product with metals such as iron, copper, or lead.

Processed Cheese: Rogers states that before the development of the can in which Cheddar cheese may be ripened, the "only commercial method for putting Cheddar cheese into a more attractive and convenient form is the one known as processing. After the rind is removed, the cheese is ground, a small quantity of water and an emulsifier, usually sodium citrate, are added, and the mass is heated with constant stirring until it becomes fluid. The emulsion is run into forms, which in many cases are boxes lined with tinfoil, in "which it is sold.

The cheese hardens quickly and, as the wrapping adheres closely, there is no trouble from molds. Moreover, as the temperature is high enough to constitute pasteurization, most of the bacteria are killed and the enzymes destroyed, so that ripening is stopped. In this process, much of the original character of the cheese is lost; but, in spite of this objection, the advantage of the package is so great that a large part, possibly one-third, of all the cheese made in the United States is sold in this form."

Templeton and Sommer have investigated various salts that may be used as emulsifiers in processed cheese. They state the purpose of the salt is to prevent separation of the fat from the cheese and at the same time give the finished product the desired body and texture. They quote Habicht as stating that an alkaline monovalent cation combined with a polyvalent anion, such as sodium citrate, is the ideal emulsifying salt. The physico-chemical explanation is as follows: There is partial saponification between the cation (sodium, if sodium citrate is used) and the fatty acids. The soaps formed are good emulsifiers. In addition the anion, which is a solvent for casein, combines with the casein of cheese so that a film of casein surrounds each fat globule, thus emulsifying it and preventing its escape from the mass. Later we find that the citrate ion is also a good peptizer of egg and flour proteins.

Loaf Cheese: Rogers states that blending is used extensively for Cheddar and Swiss cheese. In this process the cheese is ground and heated in steam-jacketed kettles, 60° to 70°, and then poured into molds. In the initial heating separation of the fat occurs; but with longer heating the casein becomes plastic and stringy and encloses the fat. Further agitation causes the mass to lose its plasticity and become the consistency of heavy cream. At this time it is poured into the molds.

The plasticity of the cheese is an important part of the process. Once the plasticity is broken it is almost impossible to restore it. The method of manufacture, the degree of ripening, the acidity of the cheese, and possibly other factors influence the degree of plasticity attainable in the heated cheese and the length of time the mass will remain plastic. Sodium and ammonia seem important in the emulsification of the product.

Cheese Spreads: The term cheese spread may be applied to any packaged form of cheese that can be easily spread with a knife at ordinary room temperature. Templeton and Sommer name the types on the market as: (1) cream cheese, mixed with pickles, olives, etc., (2) processed cheese of such age and moisture content as to be "spready," and (3) processed cheese with concentrated whey or skim milk powder added and of such fat and moisture content that the mix will spread easily. They say that, since the composition is quite different from cheese, as defined for Food and Drug regulations, the product cannot be sold as cheese. Actually they are sold as food products under proprietary trade names. The desirable spreading qualities may be due to the moisture content or the fat content or both.

The use of cheese in cooked products. All of the factors that affect the plasticity of the cheese when heated for blending, i.e., the degree of ripening, the acidity, and method of manufacture, also affect its blending properties with other ingredients in such dishes as rarebit, cheese souffle, and macaroni and cheese. To these factors may be added the extent of drying. For the cut or grated surface of cheese may dry rather extensively. Hence, the protein in the surface area really needs soaking for hydration before it will blend with other ingredients, or it may entirely lose its plasticity.

Cream cheese may be combined with eggs, sugar, etc., for cheese cake or similar cooked dishes. But, in general, whether in its original state after curing, or processed, the Cheddar type is the cheese usually combined with cooked products.

The cheese is combined with white sauce or eggs at low temperatures and by stirring. The temperature should be as low or lower than that used for blending, 40° to 50°C. often being preferable to 60° to 70°C. As the protein becomes plastic the fat exudes. Stirring aids in emulsifying this fat with white sauce and casein of the cheese.

Cheese Souffle: A colleague, Plagge, suggested a good method of combining the ingredients for cheese souffle. The beaten egg yolks were added to the white sauce before the grated cheese, because the addition of the egg yolks cooled the mixture to a greater extent before the cheese was added. However, another advantage of this order of mixing is that the egg yolk aids in emulsifying the fat of the cheese. For the same reason beating with a rotary egg beater as the cheese softens is a good method of blending the cheese with the white sauce, since it more efficiently divides the cheese, thus increasing the surface area for emulsification.

Processed cheese usually combines particularly well with white sauce and egg yolk, because of its added water content and the emulsification of the fat.

Chapter 6

Milk Microbiology

Milk is a white liquid produced by the mammary glands of mammals. It provides the primary source of nutrition for young mammals before they are able to digest other types of food. The early lactation milk is known as colostrum, and carries the mother's antibodies to the baby. It can reduce the risk of many diseases in the baby. The exact components of raw milk vary by species, but it contains significant amounts of saturated fat, protein and calcium as well as vitamin C. Cow's milk has a pH ranging from 6.4 to 6.8, making it slightly acidic.

Types of Consumption

There are two distinct types of milk consumption: a natural source of nutrition for all infant mammals and a food product for humans of all ages that is derived from other animals.

Nutrition for Infant Mammals

In almost all mammals, milk is fed to infants through breastfeeding, either directly or by expressing the milk to be stored and consumed later. Some cultures, historically or currently, continue to use breast milk to feed their children until they are seven years old.

Human infants sometimes are fed fresh goat milk. There are known risks in this practice, including those of developing electrolyte imbalances, metabolic acidosis, megaloblastic anemia, and a host of allergic reactions.

Food Product for Humans

In many cultures of the world, especially the Western world, humans continue to consume milk beyond infancy, using the milk of other animals (especially cattle, goats and sheep) as a food product. For millennia, cow's milk has been processed into dairy products such as cream, butter, yogurt, kefir, ice cream, and especially the more durable and easily transportable product, cheese. Modern industrial processes produce casein, whey protein, lactose, condensed milk, powdered milk, and many other food-additive and industrial products.

Humans are an exception in the natural world for consuming milk past infancy, despite the fact that many humans show some degree (some as little as 5%) of lactose intolerance, a characteristic that is more prevalent among individuals of African or Asian descent. The sugar lactose is found only in milk, forsythia flowers, and a few tropical shrubs.

The enzyme needed to digest lactose, lactase, reaches its highest levels in the small intestines after birth and then begins a slow decline unless milk is consumed regularly. On the other hand, those groups who do continue to tolerate milk often have exercised great creativity in using the milk of domesticated ungulates, not only of cattle, but also sheep, goats, yaks, water buffalo, horses, reindeers and camels. The largest producer and consumer of cattle and buffalo milk in the world is India.

Table: *Top ten per capita cow's milk and cow's milk products consumers in 2006*

Country	*Milk (liters)*	*Cheese (kg)*	*Butter (kg)*
Finland	183.9	19.1	5.3
Sweden	145.5	18.5	1.0
Ireland	129.8	10.5	2.9
Netherlands	122.9	20.4	3.3
Norway	116.7	16.0	4.3
Spain	119.1	9.6	1.0
Switzerland	112.5	22.2	5.6
United Kingdom	111.2	12.2	3.7
Australia	106.3	11.7	3.7
Canada	94.7	12.2	3.3

Terminology

The term *milk* is also used for white coloured, non-animal beverages resembling milk in colour and texture such as soy milk, rice milk, almond milk, and coconut milk. Even a substance regurgitated by pigeons to feed their young is called crop milk, though it bears little resemblance to mammalian milk.

Evolution

The mammary gland is thought to have been derived from apocrine skin glands. It has been suggested that the original function of lactation (milk production) was keeping eggs moist. Much of the argument is based on monotremes (egg-laying mammals): The original adaptive significance of milk secretions may have been nutrition or immunological protection.

This secretion gradually became more copious and accrued nutritional complexity over evolutionary time.

History

Animal milk is known to have been used first as human food during the Secondary Products Revolution, around 5000 BC. It is presumed that when animals such as cattle were first domesticated, it was only for purposes of meat.

Dairy products obtained from the animals proved to be a more efficient way of turning uncultivated grasslands into sustenance: the food value of an animal killed for meat can be matched by perhaps one year's worth of milk from the same animal, which will keep producing milk—in convenient daily portions—for years.

Milk by-products found inside Stone Age pottery from Turkey indicate processed milk was consumed in 6500 BC, some thousands of years before it is thought that adult humans had evolved the ability to digest raw milk. DNA evidence extracted from Neolithic skeletons indicates that in 5500 BC, people in Northern Europe, as all other peoples of the time, were still lactose intolerant. Earthenware vessels found in England and dated to 4500 BC contain milk by-products, indicating milk was used in some form, although perhaps not drunk directly.

In 1863, French chemist and biologist Louis Pasteur invented pasteurization, a method of killing harmful bacteria in beverages and food products.

In 1884, Doctor Hervey Thatcher, an American inventor from New York, invented the first glass milk bottle, called 'Thatcher's Common Sense Milk Jar', which was sealed with a waxed paper disk. Later, in 1932, plastic-coated paper milk cartons were introduced commercially as a consequence of their invention by Victor W. Farris.

The town of Harvard, Illinois celebrates milk with a summer festival known as "Milk Days". Theirs is a different tradition meant to celebrate dairy farmers in the "Milk Capital of the World."

Sources

In addition to cattle, the following livestock provide milk used by humans for dairy products:

- Camel
- Donkey
- Goat
- Horse
- Reindeer
- Sheep
- Water buffalo
- Yak.

In Russia and Sweden, small moose dairies also exist.

According to the National Bison Association, American bison (also called American buffalo) are not milked commercially, however, various sources report cows resulting from cross-breeding bison and domestic cattle are good milk producers, and have been used both during the European settlement of North America and during the development of commercial Beefalo in the 1970s and 1980s.

Human milk is not produced or distributed industrially or commercially, however, milk banks exist that allow for the collection of donated human milk and its redistribution to infants who may benefit from human milk for various reasons (premature neonates, babies with allergies, metabolic diseases, etc.)

All other female mammals do produce milk, but it rarely or never is used to produce dairy products for human consumption.

Modern Production

In the Western world today, cow's milk is produced on an industrial scale and is by far the most commonly consumed form of milk. Commercial dairy farming using automated milking equipment produces the vast majority of milk in developed countries.

Dairy cattle such as the Holstein have been bred selectively for increased milk production. About 90% of the dairy cows in the United States and 85% in Great Britain are Holsteins.

Other dairy cows in the United States include Ayrshire, Brown Swiss, Guernsey, Jersey, and Milking Shorthorn (Dairy Shorthorn).

The largest producers of dairy products and milk today are India followed by the United States, Germany, and Pakistan.

Table: *Top ten buffalo milk producers in 2007*

Country	***Production (tonnes)***	***Note***
India	59,210,000	Unofficial/Semi-official/mirror data
Pakistan	20,372,000	official figure
People's Republic of China	2,900,000	FAO estimate
Egypt	2,300,000	
Nepal	958,603	official figure
Iran	241,500	FAO estimate
Myanmar	220,462	official figure
Italy	200,000	FAO estimate
Vietnam	32,000	
Turkey	30,375	official figure
World	86,574,539	Aggregate

Increasing affluence in developing countries, as well as increased promotion of milk and milk products, has led to a rise in milk consumption in developing countries in recent years. In turn, the opportunities presented by these growing markets have attracted investment by multinational dairy firms. Nevertheless, in many countries production remains on a small scale and presents significant opportunities for diversification of income sources by small farmers.

Local milk collection centers, where milk is collected and chilled prior to being transferred to urban dairies, are a good example of where farmers have been able to work on a cooperative basis, particularly in countries such as India. The table shows the numbers for water buffalo milk production. Cattle milk is produced in a much wider range.

Grading

United States

In the United States, there are two grades of milk, with Grade A primarily used for direct sales and consumption in stores, and Grade B used for indirect consumption, such as in cheese making or other processing.

The differences between the two grades are defined in the Wisconsin administrative code for Agriculture, Trade, and Consumer Protection, chapter 60. Grade B generally refers to milk that is cooled in milk cans, which are immersed in a bath of cold flowing water, that typically is drawn up from an underground water well rather than using mechanical refrigeration.

- Grade A farms are inspected every six months, while Grade B farms are inspected every two years {WI-ATCP 60.24.2}
- Both types of farms are required to have two cleaning vats in the milk house for washing and rinsing of equipment {WI-ATCP 60.07.2 (g)}. A farm also must have an additional separate sink and faucet provided for hand washing {WI-ATCP 60.07.2(h)}, unless the bulk tank was installed before Jan 1, 1979 or the farm uses milk cans.
- Grade A milk stored in a bulk tank is cooled to 45 degrees F within two hours of milking. Grade A milk in a tank may only rise to 50 F if milk from additional milking sessions is added to the tank (potentially requiring a plate cooler to reduce the temperature of a large volume influx quickly enough) and must be cooled back to 45 F within two hours. {WI-ATCP 60.2.4(b)}
- Grade B milk in milk cans is cooled to 50 degrees F within two hours of milking. Grade B farms cannot mix milk into cans from previous milking. {WI-ATCP 60.2.4(c)}
- The somatic cell count (SCC) of Grade A or B cow or sheep milk may not exceed 750,000 cells per mL, and the SCC of Grade A or B goat milk may not exceed 1,000,000 cells per mL. {WI-ATCP 60.15.4}
- The bacterial plate or loop count of Grade A milk may not exceed 100,000 per mL, while Grade B milk may not exceed 300,000 per mL. {WI-ATCP 60.15.2}
- A bacterial plate count test is required at least once a month. {WI-ATCP 60.18.3} If the bacterial count exceeds 100,000 per mL for Grade A or 300,000 per mL for grade B in 3 out of 5 tests, the license to sell milk is suspended. The license will be revoked immediately if the bacterial count ever exceeds 750,000 per mL. {WI-ATCP 60.18.6}

Price

It was reported in 2007 that with increased worldwide prosperity and the competition of bio-fuel production for feed stocks, both the demand for and the price of milk had substantially increased world wide. Particularly notable was the rapid increase of consumption of milk in China and the rise of the price of milk in the United States above the government subsidized price.

United States

In 2010 the Department of Agriculture predicted farmers would receive an average of $1.35 per U.S. gallon of cow's milk (35 cents per liter), which is down 30 cents per gallon from 2007 and below the break-even point for many cattle farmers.

Physical and Chemical Structure

Milk is an emulsion or colloid of butterfat globules within a water-based fluid.

Butterfat

Each fat globule is surrounded by a membrane consisting of phospholipids and proteins; these emulsifiers keep the individual globules from joining together into noticeable grains of butterfat and also protect the globules from the fat-digesting activity of enzymes found in the fluid portion of the milk. In unhomogenized cow's milk, the fat globules average about four micrometers across. The fat-soluble vitamins A, D, E, and K are found within the milk fat portion of the milk.

Other Proteins

The largest structures in the fluid portion of the milk are casein protein micelles: aggregates of several thousand protein molecules, bonded with the help of nanometer-scale particles of calcium phosphate. Each micelle is roughly spherical and about a tenth of a micrometer across. There are four different types of casein proteins, and collectively they make up around 80% of the protein in milk, by weight.

Most of the casein proteins are bound into the micelles. There are several competing theories regarding the precise structure of the micelles, but they share one important feature: the outermost layer consists of strands of one type of protein, k-casein, reaching out from the body of the micelle into the surrounding fluid. These kappa-casein molecules all have a negative electrical charge and therefore repel each other, keeping the micelles separated under normal conditions and in a stable colloidal suspension in the water-based surrounding fluid.

Milk contains dozens of other types of proteins beside the caseins. They are more water-soluble than the caseins and do not form larger structures. Because these proteins remain suspended in the whey left behind when the caseins coagulate into curds, they are collectively known as *whey proteins*. Whey proteins make up approximately 20% of the protein in milk, by weight. Lactoglobulin is the most common whey protein by a large margin.

Carbohydrates

The carbohydrate lactose gives milk its sweet taste and contributes approximately 40% of whole cow's milk's calories. Lactose is a disaccharide composite of two simple sugars, glucose and galactose. In nature, lactose is found only in milk and a small number of plants. Other components found in raw cow's milk are living white blood cells, mammary gland cells, various bacteria, and a large number of active enzymes.

Appearance

Both the fat globules and the smaller casein micelles, which are just large enough to deflect light, contribute to the opaque white colour of milk. The fat globules contain some yellow-orange carotene, enough in some breeds (such as Guernsey and Jersey cattle) to impart a golden or "creamy" hue to a glass of milk. The riboflavin in the whey portion of milk has a greenish colour, which sometimes can be discerned in skimmed milk or whey products. Fat-free skimmed milk has only the casein micelles to scatter light, and they tend to scatter shorter-wavelength blue light more than they do red, giving skimmed milk a bluish tint.

Processing

In most Western countries, centralized dairy facilities process milk and products obtained from milk (dairy products), such as cream, butter, and cheese. In the U.S., these dairies usually are local companies, while in the Southern Hemisphere facilities may be run by very large nationwide or trans-national corporations (such as Fonterra).

Pasteurization

Pasteurization is used to kill harmful microorganisms by heating the milk for a short time and then cooling it for storage and transportation. Pasteurized milk still is perishable, however, and must be stored cold by both suppliers and consumers. Dairies print expiration dates on each container, after which stores will remove any unsold milk from their shelves. The process destroys the vitamin C content of the raw milk.

A newer process, ultrapasteurization or ultra-high temperature treatment (UHT), heats the milk to a higher temperature for a shorter amount of time. This extends its shelf life and allows the milk to be stored unrefrigerated because of the longer lasting sterilization effect, but it affects the taste adversely.

Microfiltration

Microfiltration is a process that partially replaces pasteurization and produces milk with fewer microorganisms and longer shelf life without a change in the taste of the milk. In this process, cream is separated from the whey and is pasteurized in the usual way, but the whey is forced through ceramic microfilters that trap 99.9% of microorganisms in the milk (as compared to 95% killing of microorganisms in conventional pasteurization). The whey then is recombined with the pasteurized cream to reconstitute the original milk composition.

Creaming and Homogenization

Upon standing for 12 to 24 hours, fresh milk has a tendency to separate into a high-fat cream layer on top of a larger, low-fat milk layer. The cream often is sold as a separate product with its own uses; today the separation of the cream from the milk usually is accomplished rapidly in centrifugal cream separators. The fat globules rise to the top of a container of milk because fat is less dense than water. The smaller the globules, the more other molecular-level forces prevent this from happening. In fact, the cream rises in cow's milk much more quickly than a simple model would predict: rather than isolated globules, the fat in the milk tends to form into clusters containing about a million globules, held together by a number of minor whey proteins. These clusters rise faster than individual globules can. The fat globules in milk from goats, sheep, and water buffalo do not form clusters so readily and are smaller to begin with; cream is very slow to separate from these milks.

Milk often is homogenized, a treatment which prevents a cream layer from separating out of the milk. The milk is pumped at high pressures through very narrow tubes, breaking up the fat globules through turbulence and cavitation. A greater number of smaller particles possess more total surface area than a smaller number of larger ones, and the original fat globule membranes cannot completely cover them. Casein micelles are attracted to the newly exposed fat surfaces; nearly one-third of the micelles in the milk end up participating in this new membrane structure. The casein weighs down the globules and interferes with the clustering that accelerated separation. The exposed fat globules are vulnerable to certain enzymes present in milk, which could break down the fats and produce rancid flavours. To prevent this, the enzymes are inactivated by pasteurizing the milk immediately before or during homogenization.

Homogenized milk tastes blander, but feels creamier in the mouth than unhomogenized; it is whiter and more resistant to developing off flavours. Creamline (or cream-top) milk is unhomogenized; it may or may not have been pasteurized. Milk which has undergone high-pressure homogenization, sometimes labelled as "ultra-homogenized," has a longer shelf life than milk which has undergone ordinary homogenization at lower pressures. Homogenized milk may be more digestible than unhomogenized milk.

Kurt A. Oster, M.D., who worked during the 1960s through the 1980s, suggested a link between homogenized milk and arterosclerosis, due to damage to plasmalogen resulting from the release of bovine xanthine oxidase (BXO) from the milk fat globular membrane (MFGM) during homogenization. Oster's hypothesis has been widely criticized, however, and has not been generally accepted by the scientific community. No link has been found between arterosclerosis and milk consumption.

Nutrition and Health

The composition of milk differs widely among species. Factors such as the type of protein; the proportion of protein, fat, and sugar; the levels of various vitamins and minerals; and the size of the butterfat globules, and the strength of the curd are among those than may vary. For example:

- Human milk contains, on average, 1.1% protein, 4.2% fat, 7.0% lactose (a sugar), and supplies 72 kcal of energy per 100 grams.
- Cow milk contains, on average, 3.4% protein, 3.6% fat, and 4.6% lactose, 0.7% minerals and supplies 66 kcal of energy per 100 grams. Donkey and horse milk have the lowest fat content, while the milk of seals and whales may contain more than 50% fat. High fat content is not unique to aquatic mammals. Guinea pig milk has an average fat content of 46%.

Table: *Milk composition analysis, per 100 grams*

Constituents	*Unit*	*Cow*	*Goat*	*Sheep*	*Waterbuffalo*
Water	g	87.8	88.9	83.0	81.1
Protein	g	3.2	3.1	5.4	4.5
Fat	g	3.9	3.5	6.0	8.0
Carbohydrate	g	4.8	4.4	5.1	4.9
Energy	kcal	66	60	95	110
Energy	kJ	275	253	396	463
Sugars (lactose)	g	4.8	4.4	5.1	4.9
Cholesterol	mg	14	10	11	8
Calcium	mg	120	100	170	195
Saturated fatty acids	g	2.4	2.3	3.8	4.2
Monounsaturated fatty acids	g	1.1	0.8	1.5	1.7
Polyunsaturated fatty acids	g	0.1	0.1	0.3	0.2

Cow's Milk

These compositions vary by breed, animal, and point in the lactation period.

Table: *Milk fat Percentages*

Cow breed	*Approximate percentage*
Jersey	5.2
Zebu	4.7
Brown Swiss	4.0
Holstein-Friesian	3.6

The protein range for these four breeds is 3.3% to 3.9%, while the lactose range is 4.7% to 4.9%. Milk fat percentages may be manipulated by dairy farmers' stock diet formulation strategies. Mastitis infection can cause fat levels to decline.

Nutritional Value

Processed cow's milk was formulated to contain differing amounts of fat during the 1950s. One cup (250 ml) of 2%-fat cow's milk contains 285 mg of calcium, which represents 22% to 29% of the daily recommended intake (DRI) of calcium for an adult.

Depending on the age, milk contains 8 grams of protein, and a number of other nutrients (either naturally or through fortification) including:

- Biotin
- Iodine
- Magnesium
- Pantothenic acid
- Potassium
- Riboflavin
- Selenium
- Thiamine
- Vitamin A
- Vitamin B_{12}
- Vitamins D
- Vitamin K.

The amount of calcium from milk that is absorbed by the human body is disputed.

Table: *Cow's milk (whole) Nutritional value per 100 g (3.5 oz)*

Energy	***252 kJ (60 kcal)***
Carbohydrates	5.26 g
Sugars	5.26 g
Lactose	5.26 g
Fat	3.25 g
saturated	1.865 g
monounsaturated	0.812 g
polyunsaturated	0.195 g
Protein	3.22 g
Tryptophan	0.075 g
Threonine	0.143 g
Isoleucine	0.165 g
Leucine	0.265 g
Lysine	0.140 g
Methionine	0.075 g
Cystine	0.017 g
Phenylalanine	0.147 g
Tyrosine	0.152 g
Valine	0.192 g
Arginine	0.075 g
Histidine	0.075 g
Alanine	0.103 g
Aspartic acid	0.237 g
Glutamic acid	0.648 g
Glycine	0.075 g
Proline	0.342 g
Serine	0.107 g
Water	88.32 g
Vitamin A equiv.	28 ìg (3%)
Thiamine (Vit. B_1)	0.044 mg (3%)
Riboflavin (Vit. B_2)	0.183 mg (12%)
Vitamin B_{12}	0.44 ìg (18%)
Vitamin D	40 IU (10%)
Calcium	113 mg (11%)
Magnesium	10 mg (3%)
Potassium	143 mg (3%)

100 mL corresponds to 103 g. Percentages are relative to US recommendations for adults. Source: USDA Nutrient database

Calcium from dairy products has a greater bioavailability than calcium from certain vegetables, such as spinach, that contain high levels of calcium-chelating agents, but a similar or lesser bioavailability than calcium from low-oxalate vegetables such as kale, broccoli, or other vegetables in the *Brassica* genus.

Medical Research

A 2006 study found that for women desiring to have a child, those who consume full fat dairy products may slightly increase their fertility, while those consuming low fat dairy products may slightly reduce their fertility. Numerous studies have found that conjugated linoleic acid, found mainly in milk, meat and dairy products, provides several health benefits including prevention of atherosclerosis, different types of cancer, and hypertension and improved immune function.

There is recent evidence suggesting consumption of milk is effective at promoting muscle growth and improving post exercise muscle recovery. In 2010, scientists at the Harvard School of Public Health identified a substance in dairy fat, trans-palmitoleic acid, that may substantially reduce the risk of type 2 diabetes. The researchers examined participants who have been followed for 20 years in an observational study to evaluate risk factors for cardiovascular diseases in older adults. During follow-up it was found that individuals with higher circulating levels of trans-palmitoleic acid had a much lower risk of developing diabetes, with about a 60% lower risk among participants in the highest quintile (fifth) of trans-palmitoleic acid levels.

Lactose Intolerance

Lactose, the disaccharide sugar component of all milk must be cleaved in the small intestine by the enzyme lactase in order for its constituents, galactose and glucose, to be absorbed. The production of this enzyme declines significantly after weaning in all mammals. Consequently, many humans become unable to digest lactose properly as they mature. There is a great deal of variance, with some individuals reacting badly to even small amounts of lactose, some able to consume moderate quantities, and some able to consume large quantities of milk and other dairy products without problems. When an individual consumes milk without producing sufficient lactase, they may suffer diarrhea, intestinal gas, cramps and bloating, as the undigested lactose travels through the gastrointestinal tract and serves as nourishment for intestinal microflora who excrete gas, a process known as anaerobic respiration.

It is estimated that 30 to 50 million Americans are lactose intolerant, including 75% of Native Americans and African Americans, and 90% of Asian Americans. Lactose intolerance is less common among those descended from northern Europeans. Lactose intolerance is a natural process and there is no reliable way to prevent or reverse it.

Controversy

Some studies suggest that milk consumption may increase the risk of suffering from certain health problems. Cow milk allergy (CMA) is an immunologically mediated adverse reaction to one or more cow's milk proteins. Rarely is it severe enough to cause death.

Milk contains casein, a substance that breaks down in the human stomach to produce casomorphin, an opioid peptide. In the early 1990s it was hypothesized that casomorphin can cause or aggravate autism spectrum disorders, and casein-free diets are widely promoted. Studies supporting these claims have had significant flaws, and the data are inadequate to guide autism treatment recommendations. A study demonstrated that men who drink a large amount of milk and consume dairy products were at a slightly increased risk of developing Parkinson's disease; the effect for women was smaller. The reason behind this is not fully understood, and it also remains unclear why there is less of a risk for women. Several sources suggest a correlation between high calcium intake (2000 mg per day, or twice the U.S. recommended daily allowance, equivalent to six or more glasses of milk per day) and prostate cancer. A large study specifically implicates dairy, i.e., low-fat milk and other dairy to which vitamin A palmitate has been added.

A review published by the World Cancer Research Fund and the American Institute for Cancer Research states that at least eleven human population studies have linked excessive dairy product consumption and prostate cancer.

Medical studies also have shown a possible link between milk consumption and the exacerbation of diseases such as Crohn's disease, Hirschsprung's disease–mimicking symptoms in babies with existing cow's milk allergies, and the aggravation of Behçet's disease.

Bovine growth Hormone Supplementation

Since November 1993, with FDA approval, Monsanto has been selling recombinant bovine somatotropin (rbST), also called rBGH, to dairy farmers. Cows produce bovine growth hormone naturally, but some producers administer an additional recombinant version of BGH which is produced through a genetically engineered E. coli because it increases milk production. Bovine growth hormone also stimulates liver production of insulin-like growth factor 1 (IGF1).

Monsanto has stated that both of these compounds are harmless given the levels found in milk and the effects of pasteurization, however, Monsanto's own tests, conducted in 1987, demonstrated that statistically significant growth stimulating effects were induced in organs of adult rats by feeding IGF-1 at low dose levels for only two weeks. "Drinking rBGH milk would thus be expected to significantly increase IGF-1 blood levels and consequently to increase risks of developing breast cancer and promoting its invasiveness."

On June 9, 2006, the largest milk processor in the world and the two largest supermarkets in the United States—Dean Foods, Wal-Mart, and Kroger—announced that they are "on a nationwide search for rBGH-free milk." Milk from cows given rBST may be sold in the United States, and the FDA stated that no significant difference has been shown between milk derived from rBST-treated and that from non-rBST-treated cows. Milk that advertises that it comes from cows not treated with rBST, is required to state this finding on its label.

Cows receiving rBGH supplements may more frequently contract an udder infection known as mastitis. Problems with mastitis have led to Canada, Australia, New Zealand, and Japan banning milk from rBST treated cows. Mastitis, among other diseases, may be responsible for the fact that levels of white blood cells in milk vary naturally.

In the European Union, rBGH is banned.

Ethical Concerns

Vegans and some other vegetarians do not consume milk for a variety of reasons. They may object to features of dairy farming including the necessity of killing almost all the male offspring of dairy cows (either by disposal soon after birth, for veal production, or for beef), the routine separation of mother and calf soon after birth, other perceived inhumane treatment of dairy cattle, and culling of cows after their productive lives.

Flavoured Milk in U.S. Schools

According to an article in *The New York Times*, milk must be offered at every meal if a United States school district wishes to get reimbursement from the federal government. A quarter of the largest school districts in the U.S. offer rice or soy milk and almost 17% of all U.S. school districts offer lactose-free milk. Seventy-one percent of the milk served in the U.S. is flavoured, causing some school districts to propose a ban because flavoured milk has added sugars. The Boulder,

Colorado school district banned flavoured milk in 2009 and instead installed a dispenser that keeps the milk colder.

Varieties and Brands

Milk products are sold in a number of varieties based on types/ degrees of;

- additives (e.g., vitamins),
- age (e.g., cheddar),
- coagulation (e.g., cottage cheese),
- farming method (e.g., organic, grass-fed).
- fat content (e.g., half and half),
- fermentation (e.g., buttermilk),
- flavoring (e.g., chocolate),
- homogenization (e.g., cream top),
- mammal (e.g., cow, goat, sheep),
- packaging (e.g., bottle),
- pasteurization (e.g., raw milk),
- water content (e.g., dry milk).

Milk preserved by the UHT process does not need to be refrigerated before opening and has a longer shelf life than milk in ordinary packaging. It is typically sold unrefrigerated in Europe, Latin America, and Australia.

Additives and Flavoring

In areas where the cattle (and often the people) live indoors, commercially sold milk commonly has vitamin D added to it to make up for lack of exposure to UVB radiation.

Reduced fat milks often have added vitamin A palmitate to compensate for the loss of the vitamin during fat removal; in the United States this results in reduced fat milks having a higher vitamin A content than whole milk.

To aid digestion in those with lactose intolerance, milk with added bacterial cultures such as *Lactobacillus acidophilus* ("acidophilus milk") and bifidobacteria ("a/B milk") is available in some areas. Another milk with *Lactococcus lactis* bacteria cultures ("cultured buttermilk") often is used in cooking to replace the traditional use of naturally soured milk, which has become rare due to the ubiquity of pasteurization, which also kills the naturally occurring lactococcus bacteria.

Milk often has flavoring added to it for better taste or as a means of improving sales. Chocolate milk has been sold for many years and has been followed more recently by strawberry milk and others. Some nutritionists have criticized flavoured milk for adding sugar, usually in the form of high-fructose corn syrup, to the diets of children who are already commonly obese in the U.S.

Distribution

Due to the short shelf-life of normal milk, it used to be delivered to households daily in many countries. However, improved refrigeration at home, changing food shopping patterns because of supermarkets, and the higher cost of home delivery, means that daily deliveries by a milkman is no longer available in most countries.

Australia and New Zealand

In Australia and New Zealand, prior to "metrification", milk was generally distributed in 1 pint (568ml) glass bottles. In Australia and in Ireland there was a government funded "free milk for school children" program, and milk was distributed at morning recess in 1/3 pint bottles. With the conversion to metric measures, the milk industry were concerned that the replacement of the pint bottles with 500ml bottles would result in a 13.6% drop in milk consumption. Hence, all pint bottles were recalled and replaced by 600ml bottles. With time, due to the steadily increasing cost of collecting, transporting, storing and cleaning glass bottles, they were replaced by cardboard cartons.

A number of designs were used, including a tetrahedron which could be close-packed without waste space, and could not be knocked over accidentally. (slogan: No more crying over spilt milk.) However, the industry eventually settled on a design similar to that used in the United States. Milk is now available in a variety of sizes in cardboard cartons (250 mL, 375 mL, 600 mL, 1 liter and 1.5 liters) and plastic bottles (1, 2 and 3 liters). A significant addition to the marketplace has been "long life" milk (UHT), generally available in 1 and 2 liter rectangular cardboard cartons. In urban and suburban areas where there is sufficient demand, home delivery is still available, though in suburban areas this is often 3 times per week rather than daily. Another significant and popular addition to the marketplace has been flavoured milks – for example, as mentioned above, Farmers Union Iced Coffee outsells Coca-Cola in South Australia.

India

In rural India, milk is delivered daily by a local milkman carrying bulk quantities in a metal container, usually on a bicycle; and in other parts of metropolitan India, milk is usually bought or delivered in a plastic bags or cartons via shops or supermarkets.

Pakistan

In Pakistan, milk is supplied in jugs. Milk has been a staple food, especially among the pastoral tribes in this country.

United Kingdom

Since the late 1990s, milk buying patterns have changed drastically in the UK. The classic milkman, who travels his local milk round (route) using a milk float (often battery powered) during the early hours and delivers milk in 1 pint glass bottles with aluminium foil tops directly to households, has almost disappeared. The main reasons for the decline of UK home deliveries by milkmen are household refrigerators (which lessen the need for daily milk deliveries) and private car usage (which has increased supermarket shopping).

In 1996, more than 2.5 billion liters of milk were still being delivered by milkmen, but by 2006 only 637 million liters (13% of milk consumed) was delivered by some 9,500 milkmen. By 2010, the estimated number of milkmen had dropped to 6,000. Assuming that delivery per milkman is the same as it was in 2006, this means milkmen deliveries now only account for 6–7% of all milk consumed by UK households (6.7 billion liters in 2008/2009). Almost 95% of all milk in the UK is thus sold in shops today, most of it in plastic bottles of various sizes, but some also in milk cartons. Milk is hardly ever sold in glass bottles in UK shops.

United States

In the United States, glass milk bottles have been replaced mostly with milk cartons and plastic jugs. Gallons of milk are almost always sold in jugs, while half-gallons and quarts may be found in both paper cartons and plastic jugs, and smaller sizes are almost always in cartons. Recently, milk has been sold in smaller resealable bottles made to fit in automobile cup holders. The .5 US pints (240 ml; 8.3 imp fl oz) milk carton is the traditional unit as a component of school lunches, though some companies have replaced that unit size with a plastic bottle, which is also available at retail in 6 and 12-pack size. Ultrapasturized milk which does not need refrigeration or in containers which can hold the cold for as long as eight hours at room temperature are available.

Packaging

Glass milk bottles are now rare. Most people purchase milk in bags, plastic bottles, or plastic-coated paper cartons. Ultraviolet (UV) light from fluorescent lighting can alter the flavour of milk, so many companies that once distributed milk in transparent or highly translucent containers are now using thicker materials that block the UV light. Milk comes in a variety of containers with local variants:

- *Australia and New Zealand:* Distributed in a variety of sizes, most commonly in aseptic cartons for up to 1.5 liters, and plastic screw-top bottles beyond that with the following volumes; 1.1 L, 2 L, and 3 L. 1 liter milk bags are starting to appear in supermarkets, but have not yet proved popular. Most UHT-milk is packed in 1 or 2 liter paper containers with a sealed plastic spout.
- *Brazil:* Used to be sold in cooled 1 liter bags, just like in South Africa. Today the most common form is 1 liter aseptic cartons containing UHT skimmed, semi-skimmed or whole milk, although the plastic bags are still in use for pasteurized milk. Higher grades of pasteurized milk can be found in cartons or plastic bottles. Sizes other than 1 liter are rare.
- *Canada:* 1.33 liter plastic bags (sold as 4 liters in 3 bags) are widely available in some areas (especially the Maritimes, Ontario and Quebec), although the 4 liter plastic jug has supplanted them in western Canada. Other common packaging sizes are 2 liter, 1 liter, 500 mL, and 250 mL cartons, as well as 4 liter, 1 liter, 250 mL aseptic cartons and 500 mL plastic jugs.
- *Chile:* Distributed most commonly in aseptic cartons for up to 1 liter, but smaller, snack-sized cartons are also popular. The most common flavours, besides the natural presentation, are chocolate, strawberry and vanilla.
- *China:* Sweetened milk is a drink popular with students of all ages and is often sold in small plastic bags complete with straw. Adults not wishing to drink at a banquet often drink milk served from cartons or milk tea.
- Colombia sells milk in 1 liter plastic bags.
- Croatia, Bosnia and Herzegovina, Serbia, Montenegro.

  UHT milk is sold in 500 mL and 1 L (sometimes also 200 ml) aseptic cartons. Non-UHT pasteurized milk is most commonly sold in 1 L and 1.5 L PET bottles, though in Serbia one can still find milk in plastic bags.
- *Parts of Europe:* Sizes of 500 mL, 1 liter (the most common), 1.5 liters, 2 liters and 3 liters are commonplace.
- *Finland:* Commonly sold in 1 L or 1.5 L cartons, in some places also in 2 dl and 5 dl cartons.

- *Hong Kong:* Milk is sold in glass bottles (220 mL), cartons (236 mL and 1 L), plastic jugs (2 liters) and aseptic cartons (250 mL).
- *India:* Commonly sold in 500 mL plastic bags and in bottles in some parts like in west. It is still customary to serve the milk boiled, despite pasteurization. Milk is often buffalo milk. Flavoured milk is sold in most convenience stores in waxed cardboard containers. Convenience stores also sell many varieties of milk (such as flavoured and ultra-pasteurized) in different sizes, usually in aseptic cartons.
- *Indonesia:* Usually sold in 1 liter cartons, but smaller, snack-sized cartons are available.
- *Israel:* Non-UHT milk is most commonly sold in 1 liter waxed cardboard boxes and 1 liter plastic bags. It may also be found in 0.5 L and 2 L waxed cardboard boxes, 2 L plastic jugs and 1 L plastic bottles. UHT milk is available in 1 liter (and less commonly also in 0.25 L) carton "bricks".
- *Japan:* Commonly sold in 1 liter waxed paperboard cartons. In most city centers there is also home delivery of milk in glass jugs. As seen in China, sweetened and flavoured milk drinks are commonly seen in vending machines.
- *South Korea:* Sold in cartons (180 mL, 200 mL, 500 mL 900 mL, 1 L, 1.8 L, 2.3 L), plastic jugs (1 L and 1.8 L), aseptic cartons (180 mL and 200 mL) and plastic bags (1 L).
- *Pakistan:* Milk is supplied in 500 mL Plastic bags and carried in Jugs from rural to cities and sell
- *Poland:* UHT milk is mostly sold in aseptic cartons (500 mL, 1 L, 2 L), and non-UHT in 1 L plastic bags or plastic bottles. Milk, UHT is commonly boiled, despite being pasteurized.
- *South Africa:* Commonly sold in 1 liter bags. The bag is then placed in a plastic jug and the corner cut off before the milk is poured.
- *Sweden:* Commonly sold in 0.3 L, 1 L or 1.5 L cartons and sometimes as plastic or glass milk bottles.
- *Turkey:* Commonly sold in 500 mL or 1L cartons or special plastic bottles. UHT milk is more popular. Milkmen also serve in smaller towns and villages.
- *United Kingdom:* Most stores stock imperial sizes: 1 pint (568 mL, 2 pints (1.136 L), 4 pints (2.273 L), 6 pints (3.408 L) or a combination including both metric and imperial sizes. Glass milk bottles delivered to the doorstep by the milkman are

typically pint-sized and are returned empty by the householder for repeated reuse. Milk is sold at supermarkets in either aseptic cartons or HDPE bottles. Supermarkets have also now begun to introduce milk in bags, to be poured from a proprietary jug and nozzle.

- *United States:* Commonly sold in gallon (3.78 L), half-gallon (1.89 L) and quart (0.94 L) containers of natural-colored HDPE resin, or, for sizes less than one gallon, cartons of waxed paperboard. Bottles made of opaque PET are also becoming commonplace for smaller, particularly metric, sizes such as one liter. The U.S. single-serving size is usually the half-pint (about 240 mL). Less frequently, dairies deliver milk directly to consumers, from coolers filled with glass bottles which are typically half-gallon sized and returned for reuse. Some convenience store chains in the United States (such as Kwik Trip in the Midwest) sell milk in half-gallon bags, while another rectangular cube gallon container design used for easy stacking in shipping and displaying is used by warehouse clubs such as Costco and Sam's Club, along with some Wal-Mart stores.
- *Uruguay:* Commonly sold in 1 liter bags. The bag is then placed in a plastic jug and the corner cut off before the milk is poured.

Practically everywhere, condensed milk and evaporated milk is distributed in metal cans, 250 and 125 mL paper containers and 100 and 200 mL squeeze tubes, and powdered milk (skim and whole) is distributed in boxes or bags.

Spoilage and Fermented Milk Products

When raw milk is left standing for a while, it turns "sour". This is the result of fermentation, where lactic acid bacteria ferment the lactose in the milk into lactic acid. Prolonged fermentation may render the milk unpleasant to consume. This fermentation process is exploited by the introduction of bacterial cultures (e.g. *Lactobacilli sp., Streptococcus sp., Leuconostoc sp.*, etc) to produce a variety of fermented milk products. The reduced pH from lactic acid accumulation denatures proteins and causes the milk to undergo a variety of different transformations in appearance and texture, ranging from an aggregate to smooth consistency. Some of these products include sour cream, yoghurt, cheese, buttermilk, viili, kefir, and kumis.

Pasteurization of cow's milk initially destroys any potential pathogens and increases the shelf-life, but eventually results in spoilage that makes it unsuitable for consumption. This causes it to assume an unpleasant odour, and the milk is deemed non-consumable due to unpleasant taste and an increased risk of food poisoning. In raw milk, the presence of lactic acid-producing bacteria, under suitable conditions, ferments the lactose present to lactic acid. The increasing acidity in turn prevents the growth of other organisms, or slows their growth significantly. During pasteurization however, these lactic acid bacteria are mostly destroyed.

In order to prevent spoilage, milk can be kept refrigerated and stored between 1 and 4 degrees Celsius in bulk tanks. Most milk is pasteurized by heating briefly and then refrigerated to allow transport from factory farms to local markets. The spoilage of milk can be forestalled by using ultra-high temperature (UHT) treatment; milk so treated can be stored unrefrigerated for several months until opened but has an unpleasant "cooked" taste. Condensed milk, made by removing most of the water, can be stored in cans for many years, unrefrigerated, as can evaporated milk. The most durable form of milk is powdered milk, which is produced from milk by removing almost all water. The moisture content is usually less than 5% in both drum- and spray-dried powdered milk.

Language and Culture

The importance of milk in human culture is attested to by the numerous expressions embedded in our languages, for example "the milk of human kindness". In ancient Greek mythology, the goddess Hera spilled her breast milk after refusing to feed Heracles, resulting in the Milky Way. In African and Asian developing nations, butter is traditionally made from fermented milk rather than cream. It can take several hours of churning to produce workable butter grains from fermented milk.

Holy books have also mentioned milk; the Bible contains references to the 'Land of Milk and Honey'. In the Qur'an, there is a request to wonder on milk as follows: 'And surely in the livestock there is a lesson for you, We give you to drink of that which is in their bellies from the midst of digested food and blood, pure milk palatable for the drinkers.' (16-The Honeybee, 66). The Ramadhan fast is traditionally broken with a glass of milk and dates.

The verb, "to milk" something is often used in the vernacular of many English-speaking countries as a synonym for extortion or, in less loaded terms, taking advantage of a situation where one has another person at a disadvantage, as in 'milking the situation'.

The word milk has had many slang meanings over time. In the early 17th century the word was used to mean semen, or vaginal secretions, or to masturbate oneself or someone else. In the 19th century, milk was used to describe a cheap alcoholic drink made from methylated spirits mixed with water. The word was also used to mean defraud, to be idle, to intercept telegrams addressed to someone else, and a weakling or 'milksop'. In the mid 1930s, the word was used in Australia meaning to siphon gas from a car.

Milk is sometimes referred to as moo juice in American English, while Cockney rhyming slang calls it Acker Bilk, Tom Silk, Lady in silk and Kilroy Silk. The name of the Russian Molokan religion in Russian is derived from Russian meaning "Milk" as they would drink milk on the Russian Orthodox days of fast.

Other Uses

Besides serving as a beverage or source of food, milk has been described as used by farmers and gardeners as an organic fungicide and foliage fertilizer. Diluted milk solutions have been demonstrated to provide an effective method of preventing powdery mildew on grape vines, while showing it is unlikely to harm the plant.

Microorganisms in Milk

The types of micro organisms found in milk vary considerably, and are dependent upon the specific conditions associated with a batch of milk. Bacteria, yeasts, moulds, and bacteriophages are commonly encountered. Other viruses and Protozoa are seldom observed in milk products, except as occasional contaminants.

Bacteria

Bacteria are the most common, and probably the most numerous of microorganisms with which the dairy processing industry is concerned. They belong to four main groups: (1) cocci, usually gram positive, (2) gram positive non-sporeforming rods, (3) gram positive sporeforming rods, and (4) gram negative non-spore forming rods.

Yeasts

The yeasts most frequently encountered in milk and milk products act upon the lactose to produce acid and carbon dioxide.

Yeasts are more commonly found in raw, cream during hot weather, but are potential contaminants throughout the year.

Moulds

Moulds often grow in large concentrations and are visible as a fuzzy or fluffy growth. They are sometimes observed on the surface, of butter, old cream, khoa, or cheese. They are black, grey, green, blue or white. They discolour milk products and often produce undesir-able, at times repulsive, odours and flavours. Moulds are essential in production of the certain kinds of cheese.

Bacteriophages

Bacteriophages are particularly obnoxious in starter cultures used for making cultured milk, butter and cheese.

Phages will, kill the bacterial culture and the whole process of fermentation will be lost.

Microorganisms found in milk can also be described on the basis of the following characteristics;

1. Biochemical activities.
2. Temperature response.
3. Ability to cause infection and disease.

Biochemical Activities

If allowed to stand under condition that permit bacterial growth, raw milk of a good sanitary quality will rapidly undergo a series of chemical changes. The principal change is lactose fermentation to lactic acid. This change is brought about by acid uric lactic organisms, especially Strepotococcus lactis and certain lactobacilli. These include two distinct biochemical types, homo-and heterofermentative. In homofermentation lactic acid is the major product of lactose fermentation. Heterofermentative organisms, however, produce lactic, acetic, propionic, and some other acids, and some alcohols and gases such as CO_2 and H_2 Organisms continue to form lactic acid until the concentration of acid is itself too great for the organisms to remain live.

Microbacteria, micrococci, coliforn, etc. also ferment lactose to lactic acid and other products. Many Clostriifiul1J species and, some yeasts such as Torula lactic, and Torula cremoris ferment lactose with acid and gas production. As the acidity continues to increase and reaches a pH of 4.7, it eventually causes a precipitation of casein. Organisms capable of metabolizing lactic and other acids develop especially acid uric, yeasts and moulds.

The acidity of milk is diminished and the alkaline products of protein decomposition such as amines, ammonia and the like are produced. This is accomplished by many species of the genera Bacillus, Clostridium, Pseudomonas, Proteus and numerous other forms.

The action of microorganisms does not involve fat as readily as it does lactose and protein. Lipolysis results from the action of lipase produced by bacteria such as Pseudomonas, Achromobacter and by some yeasts and moulds. Fat is hydrolysed to glycerol and fatty acids. Some of the fatty acids, for example, butyric and caproic acid give milk products, distinctive and usually rancid, odours and flavours.

Several microorganisms also bring about certain objection able changes in the milk which may not be deleterious to health. Rapines in milk is sometimes encountered. The milk become ropy or slimy and may be pulled out into long threads. It is produced by several organisms but the most important species is Alcaligenes viscolactis. A rapid fermentation of lactose in milk is sometimes observed and is known as stormy fermentation. This is brought about by Clostridium perfringens.

The curd become torn to shreds by the vigorous fermentation and gas production. Several organisms have been isolated from milk which impart brilliant colours. Pseudomonas syncyanea imparts blue colour, pseudomonas synxantha yellow colour and Serratia marcescens red colour to the milk.

Temperature Response

Microorganisms found in milk can also be described according their optimum temperature for growth and heat resistance. This is a very practical consideration since milk is preserved by employing low temperatures to prevent changes due to microbial activi1y' and by high temperatures to reduce microbial population and destroy pathogens. All the four types of microorganisms i.e. psychrophilic, mesophilic, thermophilic and thermoduric are found in milk.

Psychrophiles grow at temperatures just above freezing and at refrigeration tempratures. They produce a wide variety of spoilage defects. The defects may result in the production of many "off" flavours and odours. The most commonly encountered psychrophilic bacteria are members of the genera pseudomonas, Achromobacter, Vibrio, Flavobacterium and Alcaligenes, They arc killed in the pasteurization process, but are sometimes found in pasteurized milk. The contamination takes place after pasteurization from equipment, cans, bottles, and water.

The most important mesophilic bacteria are streptococci, lactobacilli and coliforms, which produce acid and gas and off flavours. They are killed in the pasteurization process

Thermophilic bacteria grow well at the temperature used in pasteurization, specially when the low temperature holding method is followed. Most thermophilic forms are found in two genera, Bacillus and Clostridium.

Thermoduric organisms are regarded as those which survive pasteurization but do not grow at pasteurization temperatures. The most common thermoduric bacteria are found in the genera Microbacterium, Corynebacterium, Micrococcus, Streptococcus and Bacillus.

Excessive numbers of thermoduric bacteria in milk make it difficult to meet the grading standard.

Ability to Cause Infection and Disease

Pathogenic organisms of both bovine and human origin have- been isolated from milk. Milk, therefore, can serve as a carrier of diseases. Many serious epidemics were caused by the consumption of such products before this fact was clearly recognized. However, this became less common as milk sanitation has improved and pasteurization is being more widely practised. The disease organisms present in milk may be derived from (1) diseased animals or (2) persons collecting and handling milk: Thus the danger is due to the inoculum and not to the growth of organisms in the milk.

The health of animal is an important factor. Several diseases of cattle including staphylococcal and streptococcal infections, tuberculosis, brucellosis, salmonellosis, Q fever and Foot and mouth disease may be transmitted to man. The organisms causing these diseases may get into the milk either directly from the udder, or indirectly from infected body discharges, which may drop, splash, or be blown into the milk. Some of the important diseases of human origin that have been transmitted by milk are (1) typhoid fever (2) diphtheria, (3) scarlet fever, (4) dysentery (5) septic sore throat and (6) poliomyelitis.

It is also possible for humans to infect animals. For example, mastitis may be caused by a variety of organisms, including Staphylococcus aureus. The infecting organism, in same cases, has been traced to humans.

Microbiology of Raw Milk

Raw milk is milk that has not been pasteurized or homogenized.

Humans consumed raw milk exclusively prior to the industrial revolution and the discovery of the pasteurization process in 1864. During the industrial revolution large populations congregated into urban areas detached from the agricultural lifestyle. Up until that point, individuals and families owned their own goats, cows, and other livestock and milked them on a daily basis.

Pasteurization was first used in the United States in the 1890s after the discovery of germ theory to control the hazards of highly contagious bacterial diseases including bovine tuberculosis and brucellosis that was thought to be easily transmitted to humans through the drinking of raw milk. Initially after the scientific discovery of bacteria, no product testing was available to determine if a farmer's milk was safe or infected, so all milk was treated as potentially contagious. After the first test was developed, some farmers actively worked to prevent their infected animals from being killed and removed from food production, or would falsify the test results so that their animals would appear to be free of infection.

When it was first used, pasteurization was thought to make raw milk from any source safer to consume. More recently, farm sanitation has greatly improved and effective testing has been developed for bovine tuberculosis and other diseases, making other approaches to ensuring safety of milk more feasible; however pasteurization continues to be widely used to prevent infected milk from entering the food supply. The recognition of many potentially deadly pathogens, such as E. Coli O157:H7, Listeria, and Salmonella, and their presence in milk products has led to the continuation of pasteurization. The Department of Health and Human Services, Centre for Disease Control and Prevention, and other health agencies of the United States strongly recommend that the public does not consume raw milk or raw milk products. Young children, the elderly, people with weakened immune systems, and pregnant women are particularly susceptible to infections originating in raw milk.

Recent advances in the analysis of milk-borne diseases have enabled scientists to track the DNA of the infectious bacteria to the cows on the farms that supplied the raw milk . This technique eliminates much of the debate of the merits of safe milk practices.

Raw vs. Pasteurized Debate

The raw vs. pasteurized debate pits the alleged health benefits of consuming raw milk against the disease threat of unpasteurized milk. Although agencies such as the Centers for Disease Control (CDC), the Food and Drug Administration (FDA), and some regulatory agencies around the world say that pathogens from raw milk make it unsafe to consume, some organizations say that raw milk can be produced hygienically, and that it has health benefits that are destroyed in the pasteurization process. This latter statement is not supported by all research.

Milk Spoilage

Spoilage occurs when microorganisms degrade the carbohydrates, proteins, fats of milk and produce noxious, end products.

It may be seen that Lactobacillus or Streptococcus species ferment the lactose to lactic acid and acetic acids turning the milk sour. They may produce enough acid to curdle the protein and form sour curd.

Attack of milk protein by Micrococcus, Bacillus or Proteus results into sweet curdling. There is little acid formation. If milk becomes contaminated with Gram-negative rods-of coliform group of bacteria, such as E. coli or Enterobacter aerogenes, or Clostridium sp., there is .acid and gas formation from the lactose.

This stormy fermentation causes the explosion of curds.

Ropiness, like bread develops from Alcaligenes, Klebsiella and Enterobacter.

Serratia marcescens causes the development of a red pigment.

Milk-borne Diseases

The important diseases are, tuberculosis, brucellosis and Q fever. Tuberculosis bacterium. Mycobacterium bovis is consumed in the milk and passes from human intestine to the blood, from which it spreads to most organs.

Brucellosis, a blood-disease is caused by Gram-negative rod, Brucella abortus. When transmitted to man through cow milk, the bacterium infect the blood-rich organs. Q fever, caused by rickettsia, Cxiella burnetii is also a milk borne disease.

Other important disorders associated with milk are primary atypical pneumonia, toxoplasmosis, anthrax, streptococcal infections etc.

Milk Microorganisms Ability to Cause Infection and Diseases

Pathogenic organisms of both bovine and human origin have- been isolated from milk. Milk, therefore, can serve as a carrier of diseases.

Many serious epidemics were caused by the consumption of such products before this fact was clearly recognized.

However, this became less common as milk sanitation has improved and pasteurization is being more widely practised.

The disease organisms present in milk may be derived from;

(1) diseased animals or

(2) persons collecting and handling milk:

Thus the danger is due to the inoculum and not to the growth of organisms in the milk.

The health of animal is an important factor. Several diseases of cattle including staphylococcal and streptococcal infections, tuberculosis, brucellosis, salmonellosis,

Q fever and Foot and mouth disease may be transmitted to man.

The organisms causing these diseases may get into the milk either directly from the udder, or indirectly from infected body discharges, which may drop, splash, or be blown into the milk.

Some of the important diseases of human origin that have been transmitted by milk are;

(1) typhoid fever

(2) diphtheria,

(3) scarlet fever,

(4) dysentery

(5) septic sore throat and

(6) poliomyelitis. It is also possible for humans to infect animals.

For example, mastitis may be caused by a variety of organisms, including Staphylococcus aureus.

The infecting organism, in same cases, has been traced to humans.

Milk Allergy

Milk allergy is a food allergy, an adverse immune reaction to one or more of the proteins in cow's milk and/or the milk of other animals—proteins that are normally harmless to the non-allergic individual. Milk allergy can cause anaphylaxis, a potentially life-threatening condition. Milk allergy is a separate and distinct condition from lactose intolerance. (Lactose intolerance is considered the normal state for most adults on a worldwide scale and is not typically considered to be a disease condition.)

Symptoms

The principal symptoms are gastrointestinal, dermatological and respiratory. These can translate to: skin rash, hives, vomiting, and gastric distress such as diarrhea, constipation, stomach pain or flatulence. The clinical spectrum extends to diverse disorders: anaphylactic reactions, atopic dermatitis, wheeze, infantile colic, gastroesophageal reflux (GER), oesophagitis, allergic colitis headache/ migraine, oral irritation, and constipation.

The symptoms may occur within a few minutes after exposure in *immediate reactions*, or after hours (and in some cases after several days) in *delayed reactions*.

Difference between Milk Allergy and Lactose Intolerance

Milk allergy is a food allergy, an adverse immune reaction to a food protein that is normally harmless to the non-allergic individual.

Lactose intolerance is a non-allergic food sensitivity, and comes from a lack of production of the enzyme lactase, required to digest the predominant sugar in milk. Adverse effects of lactose intolerance generally occur after much higher levels of milk consumption than do adverse effects of milk allergy.

Difference from Milk Protein Intolerance

Milk protein intolerance (MPI) is delayed reaction to a food protein that is normally harmless to the non-allergic, non-intolerant individual. Milk protein intolerance produces a non-IgE antibody and is not detected by allergy blood tests. Milk protein intolerance produces a range of symptoms very similar to milk allergy symptoms, but can also include blood and/or mucus in the stool. Treatment for milk protein intolerance is the same as for milk allergy. Milk protein intolerance is also referred to as milk soy protein intolerance (MSPI).

It is commonplace for milk or milk derivatives to be included in processed foods such as bread, crackers, cookies, cakes, prepared meats, "soy cheese", soups, gravies, crisps, margarine, and products labelled "non-dairy", such as whipped topping and creamer (non-dairy simply means less than 0.5% milk by weight).

In some cases, heating the dairy product to force an exothermic chemical reaction can denature the proteins, (i.e. baking bread, or other baked goods). Only the ingredients that are chemically reacting will denature.

Also, many processed foods that do not contain milk may be processed on equipment contaminated with dairy foods, which may cause an allergic reaction in some sensitive individuals.

Milk Avoidance and Replacement for Infants

Since milk protein may be transferred from a breastfeeding mother to an allergic infant, lactating mothers are put on a dairy elimination diet. For formula fed infants, milk substitute formulas are used to provide a complete source of nutrition. Milk substitutes include soy based formulas, hypoallergenic formulas based on partially or extensively hydrolyzed protein, and free amino acid-based formulas.

Non-milk derived amino acid-based formulas, known as amino acid formulas or elemental formulas, are considered the gold standard in the treatment of cows milk allergy.

Hydrolyzed formulas come in partially hydrolyzed and extensively hydrolyzed varieties. Partially hydrolyzed formulas (PHFs) are characterized by a larger proportion of long chain peptides and are considered more palatable. However, they are intended for milder cases and are not considered suitable for treatment of moderate to severe milk allergy or intolerance.

Extensively hydrolyzed formulas (EHFs) are composed of proteins that have been largely broken down into free amino acids and short peptides. Casein and whey are the most commonly used sources of protein in hydrolyzed formulas because of their high nutritional quality and their amino acid composition.

Soy based formula may or may not pose a risk of allergic sensitivity, as some infants who are allergic to milk may also be allergic to soy. Also soy based formula are not recommended for infants under 6 months. However, for infants with multiple allergies there are rice milk or oat milk based formulas available.

Milk Substitution for Children and Adults

There are many commercially available replacements for milk for children and adults - Rice milk, soy milk, oat milk, coconut milk and almond milk are also sometimes used as milk substitutes, but are not suitable nutrition for infants. However, special infant formula based on soy, rice, almonds or carob seeds is commercially available. Fruit juices supplemented with calcium which may provide an alternative for adults and children. If on an avoidance diet, it is important that dietary advice is taken as a replacement source of calcium may need to be found to prevent the longer term risk of calcium deficiency and osteoporosis. Very good dietary sources of calcium are for example sesame seeds, hemp seeds and some kinds of tofu.

Accidental Exposure

Treatment for accidental ingestion of milk products by allergic individuals varies depending on the sensitivity of the allergic person. Frequently medications such as an Epinephrine pen or an Antihistamine such as Diphenhydramine (Benadryl) are prescribed by an allergist in case of accidental ingestion. Milk allergy can cause anaphylaxis, a severe, life threatening allergic reaction.

Milk allergies are common in infants but are usually outgrown within the first 2–3 years of life.

Statistics

Milk allergy is the most common food allergy in early childhood. It affects somewhere between 2% and 3% of infants in developed countries, but approximately 85–90% of affected children lose clinical reactivity to milk once they surpass 3 years of age.

Between 13% and 20% of children allergic to milk are also allergic to beef.

Pasteurization and Sterilization of Milk

Pasteurization is used to kill harmful microorganisms by heating the milk for a short time and then cooling it for storage and transportation. Pasteurized milk still is perishable, however, and must be stored cold by both suppliers and consumers. Dairies print expiration dates on each container, after which stores will remove any unsold milk from their shelves. The process destroys the vitamin C content of the raw milk.

A newer process, ultrapasteurization or ultra-high temperature treatment (UHT), heats the milk to a higher temperature for a shorter amount of time. This extends its shelf life and allows the milk to be stored unrefrigerated because of the longer lasting sterilization effect, but it affects the taste adversely.

Liquid milk for consumption is mostly either pasteurized or sterilized. Pasteurization is a mild process, designed to inactivate the major pathogenic and spoilage bacteria in raw milk. Further improvements in shelf-life can be obtained by careful control of post-pasteurization contamination (PPC), by use of good quality raw milk and manipulations of the processing conditions. Eventually, however, such milks will spoil due to survival and growth of thermoduric bacteria or any post-pasteurization contaminants.

To keep milk for longer than few days at ambient temperature, it needs to be sterilized. The traditional process involves heating milk in a sealed container in the temperature range 114-120 Celcius degree for 20-30 minutes. More recently UHT processes have been introduced. These are continuous sterilization processes and involve temperatures in excess of 135 Celcius degree for times of greater than 1s, followed by aseptic packaging.

One of the main purposes of heat treatment is to reduce the microbial population in raw milk. Also, when milk is heated enzymes are inactivated, chemical reactions take place and there are changes in physical properties. Some important ones are a decrease in pH, precipitation of calcium phosphate, denaturation of whey proteins and interaction with casein, Maillard browning and modifications to the casein micelle.

The two most important kinetic parameters are the rate of reaction or inactivation at a constant temperature and the effect of temperature change on reaction rate. The heat resistance of vegetative bacteria and microbial spores at a constant temperatures is characterized by their decimal reduction time (D value), this is the time required to reduce the population of 90% or one decimal reduction (one log cycle). The number of decimal reductions (log N0/N) can be evaluated from:

log (N0/N) = heating time/D

where N0 = the initial population, N = final population

The two important points follow from this. Firstly, it is not possible to achieve 100% reduction. Secondly, for a specified heat treatment, the final population will increase as the initial population increases.

Pasteurization

The first stage in the history of pasteurization between 1857 and the end of the nineteenth century might well be called the medical stage, as the main history in heat-treating milk came chiefly from the medical profession interested in infant feeding. In 1927, North and Park established a wide range of temperature-time conditions to inactivate tubercle bacillus.

HTST (high temperature-short time) continuous processes were developed between 1920 and 1927 and for some time the ability of the HSTS process to produce safe milk was questioned. One method of pasteurization produces as good bottle of pasteurized milk as does the other when good methods are used and when conditions are comparable. These included test of the following:

1. Raw milk quality (platform test)
2. Pasteurizability (survival of thermodurics)
3. Efficiency of pasteurization (pathogens and phosphatase)
4. Recontamination (thermophilic and coliform bacteria and the methylene blue test)
5. General bacteria quality, including organisms surviving pasteurization plus contaminating organisms (plate court).

Enzymes in raw milk may give rise to problems in pasteurized milk. However, it is unlikely that bacterial lipases and proteases, which are very heat resistant, will cause problems in pasteurized milks because of their relatively short shelf-life and refrigerated storage conditions.

In general, the lower the storage temperature, the better is the keeping quality. Raw milk is typically stored at 4 Celcius degree, temperatures in the cold chain are slightly higher and they are likely to be higher still in domestic refrigerators. There is a requirement to further increase the shelf-life of pasteurized products, both for the convenience of the consumers and to provide additional protection against temperature abuse. However it is important to avoid the onset of cooked flavour, which would result from more severe pasteurization temperatures.

Sterilization

Sterilization of milk become a commercial proposition in 1894. Milk can be sterilized either in bottles or other sealed containers or by using ultra-high temperature (UHT) processing, which involves continuous sterilization followed by aseptic packaging. Foods have been sterilized in sealed containers, such as cans, for over 200 years. Milk was originally sterilized in glass bottles sealed with a crown cork but more recently plastic bottles are used. The main aim is to inactivate heat-resistant spores, thereby producing commercially sterile product with an extended shelf-life.

Ultra-high temperature (UHT) offers some distinct advantages over in-container sterilization. Chemical reactions are less temperature sensitive so the use of higher temperatures, combined with more rapid heating and cooling rates, helps to reduce the amount of chemical reaction. There is also a choice of indirect heat exchangers for milk, such as plate or tubular types, as well as direct steam injection or infusion plants, all of which heat products at different rates and shear conditions.

For extended shelf-life and UHT products, aseptic packaging should be used of which a number are available. They are involve putting a sterile product into a sterile container in an aseptic environment. Superheated steam has been used for sterilization of cans. Irradiation may be used for plastic bags.

Package should be inspected regularly to ensure that they are air-tight, again focusing upon those more critical part of the process. Sterilization procedures should be verified. The seal integrity of the package should be monitored as well as the overall microbial quality of packaging material itself. Rinsing, cleaning, and disinfecting procedures are also very important.

Concentrated Milks

Concentrated milk is the milk in which solids have been concentrated by removal of water. Water may be removed by evaporation, by sublimation or by partial freezing followed by removal of ice crystals.

The first method is the most common. The higher concentration of solids which results, provides the product some protection against spoilage by certain chemical actions or microbial agents. The degree of this protection is closely related to the extent of concentration; the greater the proportion of the water removed, the more the protection.

Whole milk, skim, or standardized milk are used in the preparation of concentrated milks. Partially condensed milk or skim, sweetened or unsweetened condensed milk or skim, evaporated milk or skim, and khoa are concentrated milks. Prevention of Food Adulteration (P.F.A) Act, Government of India uses the term "evaporated" to I4er to unsweetened condensed milk or skim.

Dried Milks and Dry Whey Products

Powdered milk is a manufactured dairy product made by evaporating milk to dryness. One purpose of drying milk is to preserve it; milk powder has a far longer shelf life than liquid milk and does not need to be refrigerated, due to its low moisture content. Another purpose is to reduce its bulk for economy of transportation. Powdered milk and dairy products include such items as dry whole milk, non-fat dry milk, dry buttermilk, dry whey products and dry dairy blends. Many dairy products exported conform to standards laid out in Codex Alimentarius.

Powdered milk is used for food and health (nutrition), and atypically also in biotechnology (saturating agent).

History and Manufacture

While Marco Polo wrote of Mongolian Tatar troops in the time of Kublai Khan carrying sun-dried skimmed milk as "a kind of paste," the first usable commercial production process for dried milk was invented by the Russian chemist M. Dirchoff in 1832. In 1855, T.S. Grimwade took a patent on a dried milk procedure, though a William Newton had patented a vacuum drying process as early as 1837.

In modern times, powdered milk is usually made by spray drying nonfat skim milk, whole milk, buttermilk or whey. Pasteurized milk is first concentrated in an evaporator to approximately 50% milk solids. The resulting concentrated milk is then sprayed into a heated chamber where the water almost instantly evaporates, leaving fine particles of powdered milk solids.

Alternatively, the milk can be dried by drum drying. Milk is applied as a thin film to the surface of a heated drum, and the dried milk solids are then scraped off. However, powdered milk made this way tends to have a cooked flavour, due to caramelization caused by greater heat exposure.

Another process is freeze drying, which preserves many nutrients in milk, compared to drum drying. The drying method and the heat treatment of the milk as it is processed alters the properties of the milk powder, such as its solubility in cold water, its flavour, and its bulk density.

Food and Health Uses

Powdered milk is frequently used in the manufacture of infant formula, confectionery such as chocolate and caramel candy, and in recipes for baked goods where adding liquid milk would render the product too thin. Powdered milk is also widely used in various sweets such as the famous Indian milk balls known as Rasa-Gulla and popular Indian sweet delicacy (sprinkled with desiccated coconut) known as Chum chum (made with skim milk powder).

Powdered milk is also a common item in UN food aid supplies, fallout shelters, warehouses, and wherever fresh milk is not a viable option. It is widely used in many developing countries because of reduced transport and storage costs (reduced bulk and weight, no refrigerated vehicles). As with other dry foods, it is considered nonperishable, and is favoured by survivalists, hikers, and others requiring nonperishable, easy-to-prepare food.

Reconstituting one cup of potable fluid milk from powdered milk requires one cup of potable water and one-third cup of powdered milk.

Whey Products

Whey is the portion of milk left after cheese or casein is made. Whey is about 96% water but contains some valuable protein, much lactose (milk sugar) and various dissolved salts. Years ago, whey was dumped into streams. These streams became polluted as a result; now, whey is processed into a range of dairy products.

Demineralised whey powder is made by removing dissolved salts and water. Whey protein concentrate is made by very fine filtration of whey to concentrate the proteins into a small volume which are then dried to a soluble powder.

Whey protein hydrolysates are made by digesting the protein concentrate with enzymes, and then drying the resulting soup to a powder. Milk mineral products are rich in calcium, which is extracted from the whey and then dried.

All these products are sold as ingredients to food processing companies who use them to make food products such as custards, confectionery, crab-sticks, sports drinks, baked goods and yoghurts.

Whey is the by-product of cheese and casein manufacture, being what remains of the milk once the cheese or casein is removed. Generally 100 L of milk produces about 12 kg of cheese or about 3 kg of casein. In either case, about 87 L of whey is made as a by-product.

The annual production of whey in New Zealand is about 4 x 109 L

For decades, most of New Zealand.s whey had been used mainly as pig food or run to waste.

It has also been irrigated as liquid fertiliser. Since about 1970, the New Zealand dairy industry has increasingly processed its whey into a range of products in an attempt both to extract greater total value from the parent milk and to minimise the environmental impact of whey discharges. Economic processing of whey has been aided by the progressive amalgamation of small dairy factories into larger ones where much whey arises at one site.

Major Products Made from Whey in New Zealand

Demineralised Whey Powder

Demineralised Whey Powder is produced using similar equipment to that used to produce milk powder. Whey is first subjected to either electrodialysis and/or ion exchange (described below) to reduce the mineral content by up to 90%. The demineralised whey is then evaporated to about 58% total solids and then rapidly cooled to force as much of the lactose as possible into fine crystal form. The cooled slurry is spray dried.

The spray-dried product is about 75% lactose. Spray drying is very fast and gives insufficient time for the lactose to crystallise, so that the lactose dries as a hygroscopic glass. This makes whey powders very sticky under humid conditions. The problem is minimised by first converting as much as possible of the lactose to the á-monohydrate crystal form, which is not hygroscopic. The resulting powder is sold mainly for use in infant formula preparations.

Ion exchange employs tanks full of charged plastic beads called resin. Whey is pumped down through two resin tanks: cation exchange resin in the first tank and an identical tank of anion exchange resin. The cation exchange resins are covered with sulphonate groups which carry a negative charge at all pH values above about pH 2. This resin strips metal ions from the whey and gives up H+ ions in return. The anion exchange resin is covered in quaternary amine groups which maintain a strong positive charge. This resin picks up chloride, sulphate, phosphate, citrate, nitrate and other anions from the whey in return for OH- ions.

When the cation exchange resin becomes saturated, it is regenerated ready for reuse: a solution of H2SO4 is passed through the resin bed so that the H+ ions present in high concentration displace the Ca2+, Mg2+, Na+ and K+ ions extracted from the whey. Similarly, the anion exchange resin is regenerated by NaOH solution.

Electrodialysis is a process employing a stack of porous plastic membranes sandwiched between two electrodes. Whey is pumped through the envelope made by adjacent sheets of membrane. Water is pumped through the flat spaces between those envelopes. The applied voltage encourages cations to move though the porous membranes from the whey into the water towards the cathode. Similarly anions move from whey to water towards the anode. The passage of ions from water back into whey is impeded by the chemistry of the porous membranes used. Each second membrane carries a net negative charge to make it select for the passage of cations. The alternate membranes are anion selective.

Whey Protein Concentrate (WPC)

Whey Protein Concentrate (WPC) is produced using ultrafiltration. This is also a membrane separation, but it selects on the basis of molecular size and is driven by pressure rather than by applied electric field as in the case of electrodialysis. Ultrafiltration retains (in the liquid product termed .retentate.) any insoluble material or solutes larger than about 20 000 Da molecular weight. The rest of the whey stream passes through the membrane, driven by the applied pressure and is called .permeate..

The permeate contains most of the lactose, minerals and water from the whey. The retentate, the volume of which is about 1-4% that of the feed whey, is spray dried to a powder containing 35-85% protein as desired. WPCs are made at low to moderate temperatures so that the proteins remain in their native form and the dried product is highly soluble. New Zealand is now the largest manufacturer of WPC in the world, employing many tens of square kilometres of ultrafiltration membrane. WPCs are used as food product ingredients in hams, custards, confectionery, crab-sticks, cakes, infant formulae, sports drinks and formulated stock foods.

Most WPCs contain 5-7% milkfat in the dry powdered product. This fat originated in the milk and is not removed by the cream separators through which the whey passes before ultrafiltration. The most modern WPCs use either microfiltration (like ultrafiltration but using membranes with pores sized at about 200 000 Da molecular weight) or ion exchange of the proteins themselves, prior to ultrafiltration, to make a protein product almost devoid of all fat. These very high value proteins find favour in clear acid sports beverages such as those for body-builders.

Lactalbumin

Lactalbuminin the traditional name for a product made from the same whey proteins as are in WPCs, but in an insoluble form. Lactalbumin is prepared from whey by heat precipitation under acidic conditions. Here, the major whey proteins denature and aggregate into fine flocculant curd particles. These can be separated from the liquid .serum. by centrifugation, and then washed and dried. Lactalbumin is nutritionally very valuable and is used primarily for baking, for speciality foods and to fortify some pizza cheeses. However, as the protein is insoluble, it is not useful in helping a food to gel or foam or bind together.

Whey Protein Hydrolysates

Whey Protein Hydrolysates are manufactured by enzyme digesting, at controlled temperature and pH, either WPC or lactalbumin raw materials, and then filtering and spray drying the resulting solution. Whey protein hydrolysates find uses in high value specialist nutritional applications such as tube-feeding preparations or special dietary supplements. Appropriately hydrolysed proteins lose the ability to induce allergic reactions in susceptible people, and so can be used in hypoallergenic infant formulae. Proteins are chains of amino acids, in which the amine group of one amino acid is bound to the carboxylic acid group of the neighbouring amino acid by an amide bond. Proteolytic enzymes catalyse the hydrolysis of these bonds Small chains of amino acids are called peptides. In a hydrolysate, we might seek to get all the protein into peptides of two to five amino acids, with few free amino acids and no larger peptides. Interestingly, some peptides released from milk proteins can be biologically active.

Some can transport calcium from the gut into the blood during digestion, some can inhibit enzymes in the human body involved in excessive blood pressure and some can induce sleepy feelings. However, a problem is that some peptides can be very bitter on the tongue.

Choice of the right enzymes and careful control of the hydrolysis process are required to make the desired hydrolysate product.

Milk Mineral Products

Milk mineral products rich in natural milk calcium and phosphate are valuable nutritional supplements in today osteoporosis-sensitive world. These products are made by precipitation of the calcium phosphates in whey ultrafiltration permeate under suitable conditions of concentration, pH, time and temperature. The crystals that first precipitate quickly undergo solid state transitions depending on the conditions to which they are subjected. It is necessary to grow calcium phosphate particles to sufficient size to recover them in a good yield by centrifugation and filtration. Milk mineral is used as a natural. Calcium supplement in a growing range of food products including milks, yoghurts, canned milk powders, confectionary and health foods.

Demineralised Permeate Powders

Demineralised Permeate Powders are manufactured in a very similar manner to demineralised whey powders, but use ultrafiltration permeate of whey as a raw material. The spray-dried product contains about 95% lactose.

Alcohol

The lactose in whey can be converted by fermentation by a variety of organisms to products ranging from lactic acid to flavouring materials. Three plants in New Zealand use yeast to ferment the whey ultrafiltration permeate or lactalbumin serum to ethanol. The ethanol is recovered by distillation to yield potable or industrial grade alcohol.

Chapter 7

Cream Microbiology

Cream is a dairy product that is composed of the higher-butterfat layer skimmed from the top of milk before homogenization. In un-homogenized milk, over time, the lighter fat rises to the top. In the industrial production of cream this process is accelerated by using centrifuges called "separators". In many countries, cream is sold in several grades depending on the total butterfat content. Cream can be dried to a powder for shipment to distant markets.

Cream skimmed from milk may be called "sweet cream" to distinguish it from whey cream skimmed from whey, a by-product of cheese-making. Whey cream has a lower fat content and tastes more salty, tangy and "cheesy".

Cream produced by cows (particularly Jersey cattle) grazing on natural pasture often contains some natural carotenoid pigments derived from the plants they eat; this gives the cream a slight yellow tone, hence the name of the yellowish-white colour, cream. Cream from goat's milk, or from cows fed indoors on grain or grain-based pellets, is white.

Types

Different grades of cream are distinguished by their fat content, whether they have been heat-treated, whipped, etc. In many jurisdictions there are regulations for each type.

United States

In the United States, cream is usually sold as:

- Half and half (10.5–18% fat)

- Light, coffee, or table cream (18–30% fat)
- Medium cream (25% fat)
- Whipping or light Whipping cream (30–36% fat)
- Heavy Whipping cream (36% or more)
- Extra-heavy, double, or manufacturer's cream (38–40% or more), generally not available at retail except at some warehouse and specialty stores.

Not all grades are defined by all jurisdictions, and the exact fat content ranges vary. The above figures are based on the Code of Federal Regulations, Title 21, Part 131 and a small sample of state regulations.

Australia

In Australia, levels of fat in cream are not regulated, therefore labels are only under the control of the manufacturers. A general guideline is as follows:

Extra light (or 'lite'): 12–12.5% fat.

Light (or 'lite'): 18–20% fat.

Pure cream: 35–56% fat, without artificial thickeners.

Thickened cream: 35–36.5% fat, with added gelatine and/or other thickeners to give the cream a creamier texture, also possibly with stabilisers to aid the consistency of whipped cream (this would be the cream to use for whipped cream, not necessarily for cooking)

Single cream: Recipes calling for 'single cream' are referring to pure or thickened cream with about 35% fat.

Double cream: 48–60% fat.

Canada

Canadian cream definitions are similar to those used in the United States, except for that of "light cream." In Canada, "light cream" is low-fat cream, with 5% or 6% fat. Another form of cream available in Canada is "cereal cream", which is approximately mid-way between 5% cream and coffee cream in fat content.

Japan

In Japan, cream sold in supermarkets is usually between 46% and 48% butterfat.

Processing and Additives

Cream may have thickening agents and stabilisers added. Thickeners include sodium alginate, carrageenan, gelatine, sodium bicarbonate, tetrasodium pyrophosphate, and alginic acid). Other processing may be carried out. For example, cream has a tendency to produce oily globules (called "feathering") when added to coffee. The stability of the cream may be increased by increasing the non-fat solids content, which can be done by partial demineralisation and addition of sodium caseinate, although this is expensive.

Other Cream Products

Butter is made by churning cream to separate the butterfat and buttermilk. This can be done by hand or by machine.

Whipped cream is made by whisking or mixing air into cream with more than 30% fat, to turn the liquid cream into a soft solid. Nitrous oxide may also be used to make whipped cream.

Sour cream, common in many countries including the U.S. and Australia, is cream (12 to 16% or more milk fat) that has been subjected to a bacterial culture that produces lactic acid (0.5%+), which sours and thickens it.

Crème fraîche (28% milk fat) is slightly soured with bacterial culture, but not as sour or as thick as sour cream. Mexican crema (or cream espesa) is similar to crème fraîche.

Smetana is a heavy cream product (35-40% milk fat) Central and Eastern European sour cream.

Rjome or rømme is Norwegian sour cream containing 35% milk fat, similar to Icelandic rjómi.

Clotted cream, common in the United Kingdom, is cream that has been slowly heated to dry and thicken it, producing a very high-fat (55%) product. This is similar to Indian malai.

Cream as an Ingredient

Cream is used as an ingredient in many foods, including ice cream, many sauces, soups, stews, puddings, and some custard bases, and is also used for cakes. Irish cream is an alcoholic liqueur which blends cream with whiskey and coffee. Cream is also used in curries such as masala dishes.

Cream (usually light/single cream or half and half) is often added to coffee.

For cooking purposes, both single and double cream can be used in cooking, although the former can separate when heated, usually if there is a high acid content. Most UK chefs always use double cream or full-fat crème fraîche when cream is added to a hot sauce, to prevent any problem with it separating or "splitting". In sweet and savoury custards such as those found in flan fillings, crème brûlées and crème caramels, both types of cream are called for in different recipes depending on how rich a result is called for. It is useful to note that double cream can also be thinned down with water to make an approximation of single cream if necessary.

Other Items Called "Cream"

Many non-edible substances are called creams due merely to their consistency: shoe cream is runny, unlike waxy shoe polish; face cream is a cosmetic. There is generally no restriction on describing non-edible products as creams. Regulations in many jurisdictions restrict the use of the word *cream* for foods. Words such as *creme*, *kreme*, *creame*, or *whipped topping* are often used for products which cannot legally be called cream. In some cases foods can be described as cream although they do not contain predominantly milk fats; for example in Britain "ice cream" does not have to be a dairy product (although it must be labelled "contains non-milk fat"), and salad cream is the customary name for a condiment that has been produced since the 1920s and need contain no cream.

Butter

Butter is a dairy product made by churning fresh or fermented cream or milk. It is generally used as a spread and a condiment, as well as in cooking applications, such as baking, sauce making, and pan frying. Butter consists of butterfat, water and milk proteins.

Most frequently made from cows' milk, butter can also be manufactured from the milk of other mammals, including sheep, goats, buffalo, and yaks. Salt, flavourings and preservatives are sometimes added to butter. Rendering butter produces clarified butter or *ghee*, which is almost entirely butterfat.

Butter is a water-in-oil emulsion resulting from an inversion of the cream, an oil-in-water emulsion; the milk proteins are the emulsifiers. Butter remains a solid when refrigerated, but softens to a spreadable consistency at room temperature, and melts to a thin liquid consistency at 32–35 °C (90–95 °F). The density of butter is 911 g/L (56.9 lb/ft^3).

It generally has a pale yellow colour, but varies from deep yellow to nearly white. Its unmodified colour is dependent on the animals' feed and is commonly manipulated with food colorings in the commercial manufacturing process, most commonly annatto or carotene.

Etymology

The word *butter* derives (via Germanic languages) from the Latin *butyrum*, which is the latinisation of the Greek *bouturon*. This may have been a construction meaning "cow-cheese", from *bous*, "ox, cow" + *turos*, "cheese", but perhaps this is a false etymology of a Scythian word. Nevertheless, the earliest attested form of the second stem, *turos* ("cheese"), is the Mycenaean Greek *tu-ro*, written in Linear B syllabic script. The root word persists in the name butyric acid, a compound found in rancid butter and dairy products such as Parmesan cheese. Another possibility may be an extended derivation from Sanskrit *bhutari*, meaning "the enemy of evil spirits".

In general use, the term "butter" refers to the spread dairy product when unqualified by other descriptors. The word commonly is used to describe puréed vegetable or nut products such as peanut butter and almond butter. It is often applied to spread fruit products such as apple butter. Fats such as cocoa butter and shea butter that remain solid at room temperature are also known as "butters". In addition to the act of applying butter being called "to butter", non-dairy items that have a dairy butter consistency may use "butter' to call that consistency to mind, including food items such as maple butter and Witch's butter and non-food items such as baby bottom butter, hyena butter, and rock butter.

Production

Unhomogenized milk and cream contain butterfat in microscopic globules. These globules are surrounded by membranes made of phospholipids (fatty acid emulsifiers) and proteins, which prevent the fat in milk from pooling together into a single mass. Butter is produced by agitating cream, which damages these membranes and allows the milk fats to conjoin, separating from the other parts of the cream.

Variations in the production method will create butters with different consistencies, mostly due to the butterfat composition in the finished product. Butter contains fat in three separate forms: free butterfat, butterfat crystals, and undamaged fat globules. In the finished product, different proportions of these forms result in different consistencies within the butter; butters with many crystals are harder than butters dominated by free fats.

Churning produces small butter grains floating in the water-based portion of the cream. This watery liquid is called buttermilk—although the buttermilk most common today is instead a directly fermented skimmed milk. The buttermilk is drained off; sometimes more buttermilk is removed by rinsing the grains with water. Then the grains are "worked": pressed and kneaded together. When prepared manually, this is done using wooden boards called scotch hands. This consolidates the butter into a solid mass and breaks up embedded pockets of buttermilk or water into tiny droplets.

Commercial butter is about 80% butterfat and 15% water; traditionally made butter may have as little as 65% fat and 30% water. Butterfat consists of many moderate-sized, saturated hydrocarbon chain fatty acids. It is a triglyceride, an ester derived from glycerol and three fatty acid groups. Butter becomes rancid when these chains break down into smaller components, like butyric acid and diacetly. The density of butter is 0.911 g/cm^3 (0.527 oz/in^3), about the same as ice.

Types

Before modern factory butter making, cream was usually collected from several milkings and was therefore several days old and somewhat fermented by the time it was made into butter. Butter made from a fermented cream is known as cultured butter. During fermentation, the cream naturally sours as bacteria convert milk sugars into lactic acid. The fermentation process produces additional aroma compounds, including diacetly, which makes for a fuller-flavoured and more "buttery" tasting product. Today, cultured butter is usually made from pasteurized cream whose fermentation is produced by the introduction of *Lactococcus* and *Leuconostoc* bacteria. Another method for producing cultured butter, developed in the early 1970s, is to produce butter from fresh cream and then incorporate bacterial cultures and lactic acid.

Using this method, the cultured butter flavour grows as the butter is aged in cold storage. For manufacturers, this method is more efficient, since aging the cream used to make butter takes significantly more space than simply storing the finished butter product. A method to make an artificial simulation of cultured butter is to add lactic acid and flavour compounds directly to the fresh-cream butter; while this more efficient process is claimed to simulate the taste of cultured butter, the product produced is not cultured but is instead flavoured.

Dairy products are often pasteurized during production to kill pathogenic bacteria and other microbes. Butter made from pasteurized fresh cream is called sweet cream butter. Production of sweet cream butter first became common in the 19th century, with the development of refrigeration and the mechanical cream separator. Butter made from fresh or cultured unpasteurized cream is called raw cream butter. Raw cream butter has a "cleaner" cream flavour, without the cooked-milk notes that pasteurization introduces.

Throughout Continental Europe, cultured butter is preferred, while sweet cream butter dominates in the United States and the United Kingdom. Therefore, cultured butter is sometimes labelled *European-style butter* in the United States. Commercial raw cream butter is virtually unheard-of in the United States. Raw cream butter is generally only found made at home by consumers who have purchased raw whole milk directly from dairy farmers, skimmed the cream themselves, and made butter with it. It is rare in Europe as well.

Several spreadable butters have been developed; these remain softer at colder temperatures and are therefore easier to use directly out of refrigeration. Some modify the make-up of the butter's fat through chemical manipulation of the finished product, some through manipulation of the cattle's feed, and some by incorporating vegetable oils into the butter. Whipped butter, another product designed to be more spreadable, is aerated via the incorporation of nitrogen gas—normal air is not used, because doing so would encourage oxidation and rancidity.

All categories of butter are sold in both salted and unsalted forms. Either granular salt or a strong brine are added to salted butter during processing. In addition to enhanced flavour, the addition of salt acts as a preservative.

The amount of butterfat in the finished product is a vital aspect of production. In the United States, products sold as "butter" are required to contain a minimum of 80% butterfat; in practice most American butters contain only slightly more than that, averaging around 81% butterfat. European butters generally have a higher ratio, which may extend up to 85%.

Clarified butter is butter with almost all of its water and milk solids removed, leaving almost-pure butterfat. Clarified butter is made by heating butter to its melting point and then allowing it to cool off; after settling, the remaining components separate by density. At the top, whey proteins form a skin which is removed, and the resulting butterfat is then poured off from the mixture of water and casein proteins that settle to the bottom.

Ghee is clarified butter which is brought to higher temperatures of around 120 °C (250 °F) once the water has cooked off, allowing the milk solids to brown. This process flavours the ghee, and also produces antioxidants which help protect it longer from rancidity. Because of this, ghee can keep for six to eight months under normal conditions.

Cream may be skimmed from whey instead of milk, as a by-product of cheese-making. Whey butter may be made from whey cream. Whey cream and butter have a lower fat content and taste more salty, tangy and "cheesy". They are also cheaper than "sweet" cream and butter.

European Butters

There are several butters produced in Europe with Protected geographical indications, these include:

- Beurre d'Ardenne, from Belgium
- Beurre d'Isigny, from France
- Beurre Charentes-Poitou (Which also includes: Beurre des Charentes and Beurre des Deux-Sèvres under the same classification), from France
- Beurre Rose, from Luxembourg
- Mantequilla de Soria, from Spain
- Mantega de l'Alt Urgell i la Cerdanya, from Spain.

History

The earliest butter would have been from sheep or goat's milk; cattle are not thought to have been domesticated for another thousand years. An ancient method of butter making, still used today in parts of Africa and the Near East, involves a goat skin half filled with milk, and inflated with air before being sealed. The skin is then hung with ropes on a tripod of sticks, and rocked until the movement leads to the formation of butter.

In the Mediterranean climate, unclarified butter spoils quickly—unlike cheese it is not a practical method of preserving the nutrients of milk. The ancient Greeks and Romans seemed to have considered butter a food fit more for the northern barbarians. A play by the Greek comic poet Anaxandrides refers to Thracians as *boutyrophagoi*; "butter-eaters". In *Natural History*, Pliny the Elder calls butter "the most delicate of food among barbarous nations", and goes on to describe its medicinal properties. Later, the physician Galen also described butter as a medicinal agent only.

Historian and linguist Andrew Dalby says that most references to butter in ancient Near Eastern texts should more correctly be translated as ghee. Ghee is mentioned in the Periplus of the Erythraean Sea as a typical trade article around the 1st century CE Arabian Sea, and Roman geographer Strabo describes it as a commodity of Arabia and Sudan. In India, ghee has been a symbol of purity and an offering to the gods—especially Agni, the Hindu god of fire—for more than 3000 years; references to ghee's sacred nature appear numerous times in the Rig Veda, circa 1500–1200 BCE. The tale of the child Krishna stealing butter remains a popular children's story in India today. Since India's prehistory, ghee has been both a staple food and used for ceremonial purposes such as fueling holy lamps and funeral pyres.

Middle Ages

The cooler climates of northern Europe allowed butter to be stored for a longer period before it spoiled. Scandinavia has the oldest tradition in Europe of butter export trade, dating at least to the 12th century. After the fall of Rome and through much of the Middle Ages, butter was a common food across most of Europe, but one with a low reputation, and was consumed principally by peasants. Butter slowly became more accepted by the upper class, notably when the early 16th century Roman Catholic Church allowed its consumption during Lent. Bread and butter became common fare among the middle class and the English, in particular, gained a reputation for their liberal use of melted butter as a sauce with meat and vegetables.

In antiquity, butter was used for fuel in lamps as a substitute for oil. The *Butter Tower* of Rouen Cathedral was erected in the early 16th century when Archbishop Georges d'Amboise authorized the burning of butter instead of oil, which was scarce at the time, during Lent.

Across northern Europe, butter was sometimes treated in a manner unheard-of today: it was packed into barrels (firkins) and buried in peat bogs, perhaps for years. Such "bog butter" would develop a strong flavour as it aged, but remain edible, in large part because of the unique cool, airless, antiseptic and acidic environment of a peat bog. Firkins of such buried butter are a common archaeological find in Ireland; the Irish National Museum has some containing "a grayish cheese-like substance, partially hardened, not much like butter, and quite free from putrefaction." The practice was most common in Ireland in the 11th–14th centuries; it ended entirely before the 19th century.

Industrialization

Like Ireland, France became well-known for its butter, particularly in Normandy and Brittany. By the 1860s, butter had become so in demand in France that Emperor Napoleon III offered prize money for an inexpensive substitute to supplement France's inadequate butter supplies. A French chemist claimed the prize with the invention of margarine in 1869. The first margarine was beef tallow flavoured with milk and worked like butter; vegetable margarine followed after the development of hydrogenated oils around 1900.

Until the 19th century, the vast majority of butter was made by hand, on farms. The first butter factories appeared in the United States in the early 1860s, after the successful introduction of cheese factories a decade earlier. In the late 1870s, the centrifugal cream separator was introduced, marketed most successfully by Swedish engineer Carl Gustaf Patrik de Laval. This dramatically sped up the butter-making process by eliminating the slow step of letting cream naturally rise to the top of milk. Initially, whole milk was shipped to the butter factories, and the cream separation took place there. Soon, though, cream-separation technology became small and inexpensive enough to introduce an additional efficiency: the separation was accomplished on the farm, and the cream alone shipped to the factory. By 1900, more than half the butter produced in the United States was factory made; Europe followed suit shortly after.

In 1920, Otto Hunziker authored *The Butter Industry, Prepared for Factory, School and Laboratory*, a well-known text in the industry that enjoyed at least three editions (1920, 1927, 1940). As part of the efforts of the American Dairy Science Association, Professor Hunziker and others published articles regarding: causes of tallowiness (an odour defect, distinct from rancidity, a taste defect); mottles (an aesthetic issue related to uneven colour); introduced salts; the impact of creamery metals and liquids; and acidity measurement. These and other ADSA publications helped standardize practices internationally.

Per capita butter consumption declined in most western nations during the 20th century, in large part because of the rising popularity of margarine, which is less expensive and, until recent years, was perceived as being healthier. In the United States, margarine consumption overtook butter during the 1950s and it is still the case today that more margarine than butter is eaten in the U.S. and the EU.

Shape of Butter Sticks

In the United States, butter is usually produced in 4-ounce sticks, wrapped in waxed or foiled paper and sold four to a one-pound carton. This practice is believed to have originated in 1907, when Swift and Company began packaging butter in this manner for mass distribution. Due to historical differences in butter printers (the machines which cut and package butter), these sticks are commonly produced in two different shapes:

- The dominant shape east of the Rocky Mountains is the Elgin, or Eastern-pack shape, named for a dairy in Elgin, Illinois. The sticks are 4^{3}D$_{4}$ inches long and 1 ^{1}D$_{4}$ inches (121 mm × 32 mm) wide and are typically sold stacked two by two in elongated cube-shaped boxes.
- West of the Rocky Mountains, butter printers standardized on a different shape that is now referred to as the Western-pack shape. These butter sticks are 3 ^{1}D$_{8}$ inches long and 1 ^{1}D$_{2}$ inches wide (80 mm × 38 mm) and are usually sold with four sticks packed side-by-side in a flat, rectangular box.

Both sticks contain the same amount of butter, although most butter dishes are designed for Elgin-style butter sticks.

The stick's wrapper is usually marked off as eight tablespoons (120 ml/4.2 imp fl oz; 4.1 US fl oz); the actual volume of one stick is approximately nine tablespoons (130 ml/4.6 imp fl oz; 4.4 US fl oz).

World Wide

In 1997, India produced 1,470,000 metric tons (1,620,000 short tons) of butter, most of which was consumed domestically. Second in production was the United States (522,000 t/575,000 short tons), followed by France (466,000 t/514,000 short tons), Germany (442,000 t/ 487,000 short tons), and New Zealand (307,000 t/338,000 short tons).

France ranks first in per capita butter consumption with 8 kg per capita per year. In terms of absolute consumption, Germany was second after India, using 578,000 metric tons (637,000 short tons) of butter in 1997, followed by France (528,000 t/582,000 short tons), Russia (514,000 t/567,000 short tons), and the United States (505,000 t/557,000 short tons). New Zealand, Australia, and the Ukraine are among the few nations that export a significant percentage of the butter they produce.

Different varieties are found around the world. *Smen* is a spiced Moroccan clarified butter, buried in the ground and aged for months or years. Yak butter is important in Tibet; *tsampa*, barley flour mixed with yak butter, is a staple food. Butter tea is consumed in the Himalayan regions of Tibet, Bhutan, Nepal and India. It consists of tea served with intensely flavoured—or "rancid"—yak butter and salt. In African and Asian developing nations, butter is traditionally made from sour milk rather than cream. It can take several hours of churning to produce workable butter grains from fermented milk.

Storage and Cooking

Normal butter softens to a spreadable consistency around 15 °C (60 °F), well above refrigerator temperatures. The "butter compartment" found in many refrigerators may be one of the warmer sections inside, but it still leaves butter quite hard.

Until recently, many refrigerators sold in New Zealand featured a "butter conditioner", a compartment kept warmer than the rest of the refrigerator—but still cooler than room temperature—with a small heater. Keeping butter tightly wrapped delays rancidity, which is hastened by exposure to light or air, and also helps prevent it from picking up other odors. Wrapped butter has a shelf life of several months at refrigerator temperatures.

"French butter dishes" or "Acadian butter dishes" involve a lid with a long interior lip, which sits in a container holding a small amount of water. Usually the dish holds just enough water to submerge the interior lip when the dish is closed. Butter is packed into the lid. The water acts as a seal to keep the butter fresh, and also keeps the butter from overheating in hot temperatures. This allows butter to be safely stored on the countertop for several days without spoilage.

Once butter is softened, spices, herbs, or other flavoring agents can be mixed into it, producing what is called a *compound butter* or *composite butter* (sometimes also called *composed butter*). Compound butters can be used as spreads, or cooled, sliced, and placed onto hot food to melt into a sauce. Sweetened compound butters can be served with desserts; such hard sauces are often flavoured with spirits.

Melted butter plays an important role in the preparation of sauces, most obviously in French cuisine. *Beurre noisette* (hazelnut butter) and *Beurre noir* (black butter) are sauces of melted butter cooked until the milk solids and sugars have turned golden or dark brown; they are often finished with an addition of vinegar or lemon juice.

Hollandaise and béarnaise sauces are emulsions of egg yolk and melted butter; they are in essence mayonnaises made with butter instead of oil. Hollandaise and béarnaise sauces are stabilized with the powerful emulsifiers in the egg yolks, but butter itself contains enough emulsifiers—mostly remnants of the fat globule membranes—to form a stable emulsion on its own. *Beurre blanc* (white butter) is made by whisking butter into reduced vinegar or wine, forming an emulsion with the texture of thick cream. *Beurre monté* (prepared butter) is melted but still emulsified butter; it lends its name to the practice of "mounting" a sauce with butter: whisking cold butter into any water-based sauce at the end of cooking, giving the sauce a thicker body and a glossy shine—as well as a buttery taste.

In Poland, the butter lamb (*Baranek wielkanocny*) is a traditional addition to the Easter Meal for many Polish Catholics. Butter is shaped into a lamb either by hand or in a lamb-shaped mould.

Butter is also used to make edible decorations to garnish other dishes.

Butter is used for sautéing and frying, although its milk solids brown and burn above 150 °C (250 °F)—a rather low temperature for most applications. The smoke point of butterfat is around 200 °C (400 °F), so clarified butter or ghee is better suited to frying. Ghee has always been a common frying medium in India, where many avoid other animal fats for cultural or religious reasons.

Butter fills several roles in baking, where it is used in a similar manner as other solid fats like lard, suet, or shortening, but has a flavour that may better complement sweet baked goods.

Many cookie doughs and some cake batters are leavened, at least in part, by creaming butter and sugar together, which introduces air bubbles into the butter. The tiny bubbles locked within the butter expand in the heat of baking and aerate the cookie or cake. Some cookies like shortbread may have no other source of moisture but the water in the butter.

Pastries like pie dough incorporate pieces of solid fat into the dough, which become flat layers of fat when the dough is rolled out. During baking, the fat melts away, leaving a flaky texture.

Butter, because of its flavour, is a common choice for the fat in such a dough, but it can be more difficult to work with than shortening because of its low melting point. Pastry makers often chill all their ingredients and utensils while working with a butter dough.

Butter also has many non-culinary, traditional uses which are specific to certain cultures. For instance, in North America, applying butter to the handle of a door is a common prank on April Fools' Day.

Health and Nutrition

Table: *Butter, Unsalted Nutritional Value Per 100 g (3.5 oz)*

Energy	2,999 kJ (717 kcal)
Carbohydrates	0 g
Fat	81 g
Saturated	51 g
Monounsaturated	21 g
Polyunsaturated	3 g
Protein	1 g
Vitamin A equiv.	684 μg (76%)
Vitamin D	60 IU (15%)
Vitamin E	2.32 mg (15%)
Cholesterol	215 mg

Fat percentage can vary.

Percentages are relative to US recommendations for adults.

Source: USDA Nutrient database

According to USDA figures, one tablespoon of butter (14 grams / 0.5 ounces) contains 420 kilojoules (100 kcal), all from fat, 11 grams (0.4 oz) of fat, of which 7 grams (0.25 oz) are saturated fat, and 30 milligrams (0.46 gr) of cholesterol.

Butter consists mostly of saturated fat and is a significant source of cholesterol. For these reasons butter is considered by some to be a contributor to health problems, especially heart disease.

Margarine was recommended as a substitute, since it is higher in unsaturated fat and contains little or no cholesterol, but in recent years, it has been shown that the trans fats contained in partially hydrogenated oils used in typical margarines significantly raise undesirable LDL cholesterol levels as well. Trans-fat–free margarines have since been developed.

Proponents of the consumption of organic butter, such as the nutritionist Mary Enig, state that since butter is nutritious and "is rich in short and medium chain fatty acids," this can have a positive effect on health and prevent disease.

Butter contains only traces of lactose, so moderate consumption of butter is not a problem for the lactose intolerant. People with milk allergies need to avoid butter, which contains enough of the allergy-causing proteins to cause reactions.

Butter can perform a useful role in dieting by providing satiety. A small amount added to low fat foods such as vegetables may stave off feelings of hunger.

Icecream and Related Frozen Dairy Desserts

Ice cream is a frozen dessert usually made from dairy products, such as milk and cream, and often combined with fruits or other ingredients and flavours. Most varieties contain sugar, although some are made with other sweeteners. In some cases, artificial flavourings and colorings are used in addition to (or in replacement of) the natural ingredients. This mixture is stirred slowly while cooling to prevent large ice crystals from forming; the result is a smoothly textured ice cream.

The meaning of the term ice cream varies from one country to another. Terms like frozen custard, frozen yogurt, sorbet, gelato and others are used to distinguish different varieties and styles. In some countries, like the USA, the term ice cream applies only to a specific variety, and their governments regulate the commercial use of all these terms based on quantities of ingredients. In others, like Italy and Argentina, one word is used for all the variants. Alternatives made from soy milk, rice milk, and goat milk are available for those who are lactose intolerant or have an allergy to dairy protein, or in the case of soy and rice milk, for those who want to avoid animal products.

History

Precursors of Ice Cream

In the Persian Empire, people would pour grape juice concentrate over snow - in a bowl - and eat this as a treat. In particular this was consumed when the weather was hot. Either snow would be saved in the cool-keeping underground chambers known as "yakhchal" or taken from fresh snow that may still have remained at the top of the mountains by the summer capital - Hagmatana, Ecbatana or Hamedan of today. In 400 BC, the Persians went further and invented a special chilled food, made of rose water and vermicelli which was served to royalty during summers. The ice was mixed with saffron, fruits, and various other flavours.

Ancient civilizations have served ice for cold foods for thousands of years. The BBC reports that a frozen mixture of milk and rice was used in China around 200 BC. The Roman Emperor Nero (37–68) had ice brought from the mountains and combined with fruit toppings. These were some early chilled delicacies.

Arabs were the first to use milk as a major ingredient in its production, sweeten the ice cream with sugar rather than fruit juices, as well as perfect ways for its commercial production. As early as the 10th century, ice cream was widespread amongst many of the Arab world's major cities, such as Baghdad, Damascus and Cairo.

Their version of ice cream was produced from milk or cream and often some yoghurt similar to Ancient Greek recipes, flavoured with rosewater as well as dried fruits and nuts. It is believed that this was based on older Ancient Arab, Mesopotamian, Greek or Roman recipes, which were probably the first and precursors to Persian faloodeh. In 62 AD, the Roman emperor Nero sent slaves to the Apennine mountains to collect snow to be flavoured with honey and nuts.

Maguelonne Toussaint-Samat asserts in her *History of Food*, "the Chinese may be credited with inventing a device to make sorbets and ice cream. They poured a mixture of snow and saltpetre over the exteriors of containers filled with syrup, for, in the same way as salt raises the boiling-point of water, it lowers the freezing-point to below zero." (Toussaint does not provide historical documentation for this.) Some distorted accounts claim that in the age of Emperor Yingzong, Song Dynasty (960-1279) of China, was written by the poet Yang Wanli. Actually, and has nothing to do with ice cream.

It has also been claimed that, in the Yuan Dynasty, Kublai Khan enjoyed ice cream and kept it a royal secret until Marco Polo visited China and took the technique of making ice cream to Italy. Others have argued that the Chinese didn't drink milk during that period, whereas the Italians had arguably been making something resembling ice cream before Marco Polo returned to Italy. In any case, no known ice cream recipes appear to stem from ancient Chinese sources.

In the sixteenth century, the Mughal emperors used relays of horsemen to bring ice from the Hindu Kush to Delhi, where it was used in fruit sorbets.

When Italian duchess Catherine de' Medici married the duc d'Orléans in 1533, she is said to have brought with her Italian chefs who had recipes for flavoured ices or sorbets, and introduced them in France. One hundred years later, Charles I of England was supposedly so impressed by the "frozen snow", he offered his own ice cream maker a lifetime pension in return for keeping the formula secret, so ice cream could be a royal prerogative. There is no historical evidence to support these legends, which first appeared during the 19th century.

The first recipe for flavoured ices in French appears in 1674, in Nicholas Lemery's *Recueil de curiositéz rares et nouvelles de plus admirables effets de la nature*. Recipes for *sorbetti* saw publication in the 1694 edition of Antonio Latini's *Lo Scalco alla Moderna* (The Modern Steward). Recipes for flavoured ices begin to appear in François Massialot's *Nouvelle Instruction pour les Confitures, les Liqueurs, et les Fruits* starting with the 1692 edition. Massialot's recipes result in a coarse, pebbly texture. Latini claims that the results of his recipes should have the fine consistency of sugar and snow.

True Ice Cream

Ice cream recipes first appear in 18th century England and America. A recipe for ice cream was published in *Mrs. Mary Eales's Receipts* in London 1718.

To ice CREAM. Take Tin Ice-Pots, fill them with any Sort of Cream you like, either plain or sweeten'd, or Fruit in it; shut your Pots very close; to six Pots you must allow eighteen or twenty Pound of Ice, breaking the Ice very small; there will be some great Pieces, which lay at the Bottom and Top: You must have a Pail, and lay some Straw at the Bottom; then lay in your Ice, and put in amongst it a Pound of Bay-Salt; set in your Pots of Cream, and lay Ice and Salt between every Pot, that they may not touch; but the Ice must lie round them on every Side; lay a good deal of Ice on the Top, cover the Pail with Straw, set it in a Cellar where no Sun or Light comes, it will be froze in four Hours, but it may stand longer; than take it out just as you use it; hold it in your Hand and it will slip out. When you would freeze any Sort of Fruit, either Cherries, Raspberries, Currants, or Strawberries, fill your Tin-Pots with the Fruit, but as hollow as you can; put to them Lemmonade, made with Spring-Water and Lemmon-Juice sweeten'd; put enough in the Pots to make the Fruit hang together, and put them in Ice as you do Cream.

The earliest reference to ice cream given by the *Oxford English Dictionary* is from 1744, reprinted in a magazine in 1877. *1744 in Pennsylvania Mag. Hist. & Biogr. (1877) I. 126 Among the rarities..was some fine ice cream, which, with the strawberries and milk, eat most deliciously*. The 1751 edition of *The Art of Cookery made Plain and Easy* by Hannah Glasse features a recipe for ice cream. OED gives her recipe: *H. GLASSE Art of Cookery (ed. 4) 333 (heading) To make Ice Cream..set it [sc. the cream] into the larger Bason. Fill it with Ice, and a Handful of Salt.*

1768 saw the publication of *L'Art de Bien Faire les Glaces d'Office* by M. Emy, a cookbook devoted entirely to recipes for flavoured ices and ice cream.

Ice cream was introduced to the United States by Quaker colonists who brought their ice cream recipes with them. Confectioners sold ice cream at their shops in New York and other cities during the colonial era. Ben Franklin, George Washington, and Thomas Jefferson were known to have regularly eaten and served ice cream. First Lady Dolley Madison is also closely associated with the early history of ice cream in the United States. One respected history of ice cream states that, as the wife of U.S. President James Madison, she served ice cream at her husband's Inaugural Ball in 1813.

Around 1832, Augustus Jackson, an African American confectioner, not only created multiple ice cream recipes, but he also invented a superior technique to manufacture ice cream.

In 1843, Nancy Johnson of Philadelphia was issued the first U.S. patent for a small-scale handcranked ice cream freezer. The invention of the ice cream soda gave Americans a new treat, adding to ice cream's popularity. This cold treat was probably invented by Robert Green in 1874, although there is no conclusive evidence to prove his claim.

The ice cream sundae originated in the late 19th century. Several men claimed to have created the first sundae, but there is no conclusive evidence to back up any of their stories. Some sources say that the sundae was invented to circumvent blue laws, which forbade serving sodas on Sunday. Towns claiming to be the birthplace of the sundae include Buffalo, New York; Two Rivers, Wisconsin; Ithaca, New York; and Evanston, Illinois.

Both the ice cream cone and banana split became popular in the early 20th century. Several food vendors claimed to have invented the ice cream cone at the 1904 World's Fair in St. Louis, MO. Europeans were eating cones long before 1904. In the UK, ice cream remained an expensive and rare treat, until large quantities of ice began to be imported from Norway and the US in the mid Victorian era. A Swiss-Italian businessman, Carlo Gatti, opened the first ice cream stall outside Charing Cross station in 1851, selling scoops of ice cream in shells for one penny.

The history of ice cream in the 20th century is one of great change and increases in availability and popularity. In the United States in the early 20th century, the ice cream soda was a popular treat at the soda shop, the soda fountain, and the ice cream parlor. During American

Prohibition, the soda fountain to some extent replaced the outlawed alcohol establishments such as bars and saloons.

Ice cream became popular throughout the world in the second half of the 20th century after cheap refrigeration became common. There was an explosion of ice cream stores and of flavours and types. Vendors often competed on the basis of variety. Howard Johnson's restaurants advertised "a world of 28 flavours." Baskin-Robbins made its 31 flavours ("one for every day of the month") the cornerstone of its marketing strategy. The company now boasts that it has developed over 1000 varieties. One important development in the 20th century was the introduction of soft ice cream.

A chemical research team in Britain (of which a young Margaret Thatcher was a member) discovered a method of doubling the amount of air in ice cream, which allowed manufacturers to use less of the actual ingredients, thereby reducing costs. It made possible the soft ice cream machine in which a cone is filled beneath a spigot on order. In the United States, Dairy Queen, Carvel, and Tastee-Freez pioneered in establishing chains of soft-serve ice cream outlets.

Technological innovations such as these have introduced various food additives into ice cream, notably the stabilizing agent gluten, to which some people have an intolerance. Recent awareness of this issue has prompted a number of manufacturers to start producing gluten-free ice cream. The 1980s saw a return of the older, thicker ice creams being sold as "premium" and "superpremium" varieties under brands such as Ben & Jerry's and Häagen-Dazs.

Production

Before the development of modern refrigeration, ice cream was a luxury reserved for special occasions. Making it was quite laborious; ice was cut from lakes and ponds during the winter and stored in holes in the ground, or in wood-frame or brick ice houses, insulated by straw. Many farmers and plantation owners, including U.S. Presidents George Washington and Thomas Jefferson, cut and stored ice in the winter for use in the summer. Frederic Tudor of Boston turned ice harvesting and shipping into a big business, cutting ice in New England and shipping it around the world.

Ice cream was made by hand in a large bowl placed inside a tub filled with ice and salt. This was called the pot-freezer method. French confectioners refined the pot-freezer method, making ice cream in a sorbetière (a covered pail with a handle attached to the lid). In the pot-freezer method, the temperature of the ingredients is reduced by the mixture of crushed ice and salt. The salt water is cooled by the ice, and the action of the salt on the ice causes it to (partially) melt, absorbing latent heat and bringing the mixture below the freezing point of pure water. The immersed container can also make better thermal contact with the salty water and ice mixture than it could with ice alone.

The hand-cranked churn, which also uses ice and salt for cooling, replaced the pot-freezer method. The exact origin of the hand-cranked freezer is unknown, but the first U.S. patent for one was #3254 issued to Nancy Johnson on September 9, 1843. The hand-cranked churn produced smoother ice cream than the pot freezer and did it quicker. Many inventors patented improvements on Johnson's design.

In Europe and early America, ice cream was made and sold by small businesses, mostly confectioners and caterers. Jacob Fussell of Baltimore, Maryland was the first to manufacture ice cream on a large scale. Fussell bought fresh dairy products from farmers in York County, Pennsylvania, and sold them in Baltimore. An unstable demand for his dairy products often left him with a surplus of cream, which he made into ice cream. He built his first ice-cream factory in Seven Valleys, Pennsylvania, in 1851. Two years later, he moved his factory to Baltimore. Later, he opened factories in several other cities and taught the business to others, who operated their own plants. Mass production reduced the cost of ice cream and added to its popularity.

The development of industrial refrigeration by German engineer Carl von Linde during the 1870s eliminated the need to cut and store natural ice and when the continuous-process freezer was perfected in 1926, it allowed commercial mass production of ice cream and the birth of the modern ice cream industry.

The most common method for producing ice cream at home is to use an ice cream maker, in modern times generally an electrical device that churns the ice cream mixture while cooled inside a household freezer, or using a solution of pre-frozen salt and water, which gradually melts while the ice cream freezes. Some more expensive models have an inbuilt freezing element. A newer method of making home-made ice cream is to add liquid nitrogen to the mixture while stirring it using a spoon or spatula. Some ice cream recipes call for making a custard, folding in whipped cream, and immediately freezing the mixture.

Commercial Delivery

Ice cream can be mass-produced and thus is widely available in developed parts of the world. Ice cream can be purchased in large cartons (vats and squrounds) from supermarkets and grocery stores, in smaller quantities from ice cream shops, convenience stores, and milk bars, and in individual servings from small carts or vans at public events. In Turkey and Australia, ice cream is sometimes sold to beach-goers from small powerboats equipped with chest freezers. Some ice cream distributors sell ice cream products from traveling refrigerated vans or carts (commonly referred to in the US as "ice cream trucks"), sometimes equipped with speakers playing children's music. Traditionally, ice cream vans in the United Kingdom make a music box noise rather than actual music.

Dietary

Ice cream may have the following composition:

- greater than 10% milkfat and usually between 10% and as high as 16% fat in some premium ice creams
- 9 to 12% milk solids-not-fat: this component, also known as the serum solids, contains the proteins (caseins and whey proteins) and carbohydrates (lactose) found in milk
- 12 to 16% sweeteners: usually a combination of sucrose and glucose-based corn syrup sweeteners
- 0.2 to 0.5% stabilisers and emulsifiers
- 55% to 64% water which comes from the milk or other ingredients.

These compositions are percentage by weight. Since ice cream can contain as much as half air by volume, these numbers may be reduced by as much as half if cited by volume. In terms of dietary considerations, The percentages by weight are more relevant. Even the low fat products have high caloric content: Ben and Jerry's No Fat Vanilla Fudge contains 150 calories per half cup due to its high sugar content.

Ice Cream around the World

Ice Cream Cone

Mrs Marshall's Cookery Book, published in 1888, endorsed serving ice cream in cones, but the idea definitely predated that. Agnes Marshall was a celebrated cookery writer of her day and helped to popularise ice cream. She patented and manufactured an ice cream maker and was the first person to suggest using liquefied gases to freeze ice cream after seeing a demonstration at the Royal Institution.

Reliable evidence proves that ice cream cones were served in the 19th century, and their popularity increased greatly during the St. Louis World's Fair in 1904. According to legend, at the World's Fair an ice cream seller had run out of the cardboard dishes used to put ice cream scoops in, so they could not sell any more produce.

Next door to the ice cream booth was a Syrian waffle booth, unsuccessful due to intense heat; the waffle maker offered to make cones by rolling up his waffles and the new product sold well, and was widely copied by other vendors.

Other Frozen Desserts

The following is a partial list of ice cream-like frozen desserts and snacks:

- Ais kacang: a dessert in Malaysia and Singapore made from shaved ice, syrup, and boiled red bean and topped with evaporated milk. Sometimes, other small ingredients like raspberries and durians are added in too.
- Dondurma: Turkish ice cream, made of salep and mastic resin
- Frozen custard: at least 10% milk fat and at least 1.4% egg yolk and much less air beaten into it, similar to Gelato, fairly rare. Known in Italy as Semifreddo.
- Frozen yogurt: a low fat or fat free alternative made with yogurt
- Gelato: an Italian frozen dessert having a lower milk fat content than ice cream and stabilised with ingredients such as eggs.
- Halo-halo: a popular Filipino dessert that is a mixture of shaved ice and milk to which are added various boiled sweet beans and fruits, and served cold in a tall glass or bowl.
- Ice milk: less than 10% milk fat and lower sweetening content, once marketed as "ice milk" but now sold as *low-fat ice cream* in the United States.
- Ice pop (or lolly): frozen fruit puree, fruit juice, or flavoured sugar water on a stick or in a flexible plastic sleeve.
- Kulfi: Believed to have been introduced to South Asia by the Mughal conquest in the 16th century; its origins trace back to the cold snacks and desserts of Arab and Mediterranean cultures.
- Mellorine: non-dairy, with vegetable fat substituted for milk fat
- Parevine: Kosher non-dairy frozen dessert established in 1969 in New York
- Sherbet: 1–2% milk fat and sweeter than ice cream.

- Sorbet: fruit puree with no dairy products
- Snow cones, made from balls of crushed ice topped with sweet syrup served in a paper cone, are consumed in many parts of the world. The most common places to find snow cones in the United States are at amusement parks.
- Maple toffee: A popular springtime treat in maple-growing areas is maple toffee, where maple syrup boiled to a concentrated state is poured over fresh snow congealing in a toffee-like mass, and then eaten from a wooden stick used to pick it up from the snow.

Using Liquid Nitrogen

Using liquid nitrogen to freeze ice cream is an old idea and has been used for many years to harden ice cream. The use of liquid nitrogen in the primary freezing of ice cream, that is to effect the transition from the liquid to the frozen state without the use of a conventional ice cream freezer, has only recently started to see commercialization. Some commercial innovations have been documented in the National Cryogenic Society Magazine "Cold Facts". The most noted brands are Dippin' Dots, Blue Sky Creamery, Project Creamery, and Sub Zero Cryo Creamery. The preparation results in a column of white condensed water vapor cloud, reminiscent of popular depictions of witches' cauldrons. The ice cream, dangerous to eat while still "steaming," is allowed to rest until the liquid nitrogen is completely vaporised. Sometimes ice cream is frozen to the sides of the container, and must be allowed to thaw.

Making ice cream with liquid nitrogen has advantages over conventional freezing. Due to the rapid freezing, the crystal grains are smaller, giving the ice cream a creamier texture, and allowing one to get the same texture by using less milkfat. Such ice crystals will grow very quickly via the processes of recrystallization thus obviating the original benefits unless steps are taken to inhibit ice crystal growth.

Gelato

Gelato is Italy's regional variant of ice cream. As such, gelato is made with some of the same ingredients as most other frozen dairy desserts. Milk, cream, various sugars, flavoring including fruit and nut purees are the main ingredients.

Gelato is different from some other ice creams because it has a lower butterfat content. Gelato typically contains 4-8% butterfat, versus 14% for many other ice creams. Gelato generally has slightly lower sugar content, averaging between 16-22% versus approximately 21% for most ice creams. Non-fat milk is added as a solid. The sugar content in gelato is precisely balanced with the water content to act as an anti-freeze to prevent the gelato from freezing solid. Types of sugar used include sucrose, dextrose, and invert sugar to control apparent sweetness. Typically, gelato and Italian sorbet contain a stabilizing base. Egg yolks are used in yellow custard-based gelato flavours, including zabaione and creme caramel.

The mixture for gelato is typically made using a hot process, which includes pasteurization. White base is heated to 85°C (185°F). Heating the mix to 90°C (194°F) is essential for chocolate gelato, which is traditionally flavoured with cocoa powder. Yellow custard base, which contains egg yolks, is heated to 65°C (149°F).

The gelato mix must age for several hours after pasteurization is complete for the milk proteins to hydrate, or bind, with water in the mix. This hydration reduces the size of the ice crystals, making a smoother texture in the final product. A non-traditional cold mix process is popular among some gelato makers in the United States.

Unlike most commercial ice creams in the United States, which are frozen with a continuous assembly line freezer, gelato is frozen very quickly in individual small batches in a batch freezer. The batch freezer incorporates air or overage into the mix as it freezes. Unlike most American-style ice creams, which can have an overage of up to 50%, gelato generally has between 20% and 35% overage. This results in a denser product with more intense flavour than many U.S. style ice creams. U.S. style ice cream, with a higher fat content, can be stored in a freezer for months. High-quality artisan gelato holds its peak flavour and texture (from delicate ice crystals) only for several days, even when stored carefully at the proper temperature. This is why *gelaterias* typically make their own gelato on the premises or nearby.

History

The history of gelato dates back to frozen desserts served in ancient Rome and Egypt made from ice and snow brought down from mountaintops and preserved below ground. Later, gelato appeared during banquets at the Medici court in Florence. In fact, the Florentine cook Bernardo Buontalenti is said to have invented modern ice cream in 1565, as he presented his recipe and his innovative refrigerating techniques to Catherine de' Medici. She in turn brought the novelty to France, where in 1686 the Sicilian fisherman Francesco Procopio dei Coltelli perfected the first ice cream machine. The popularity of gelato among larger shares of the population however only increased in the 1920s-1930s as in the northern Italian city of Varese, where the first mobile gelato cart was developed.

Gelato is typically flavoured with fresh fruit purees, cocoa and/or nut pastes. If other ingredients such as chocolate flakes, nuts, small confections, cookies, or biscuits are added, they are added after the gelato is frozen. Gelato made with fresh fruit, sugar, and water and without dairy ingredients is sorbet. Use of the word sorbetto is a common affectation in the United States. In Italy, the word sorbetto refers to a cocktail.

Stracciatella

Stracciatella (from Italian *stracciato*, "torn apart") is an Italian egg-drop soup usually said to be "alla Romana" ("the way it's done in Rome"), but also popular in Marche and Emilia Romagna. It is prepared by beating eggs and adding grated parmesan cheese, salt, pepper, nutmeg, and sometimes semolina, and then adding this mixture to boiling broth. The broth is set whirling first with a whisk, and the beaten egg mixture added in a slow stream to produce the *stracciatelle* ("little shreds") of cooked egg in the broth, which is clarified by the process.

In Gelato

In Italy and Germany, gelato with a vanilla base and chocolate shavings is also called stracciatella or stracciatella ice. It is somewhat analogous to chocolate chip ice cream in North America though the chocolate is intended to be less chunky and more integrated with the gelato.

In Cheese

A particular kind of mozzarella (soft cheese) is also called stracciatella. Stracciatella is used as stuffing for the burrata from the Murgia region in Puglia. It is made with torn pieces of mozzarella and cream.

Frozen Custard

Frozen custard is a cold dessert similar to ice cream, made with eggs in addition to cream and sugar.

In the United States the Food and Drug Administration requires products marketed as frozen custard to contain at least 10 percent milkfat and 1.4 percent egg yolk solids. If it has less egg yolk solids, it is considered ice cream. In the United Kingdom frozen custard is not differentiated from other frozen desserts. Instead, if given a name other than frozen desserts they may be referred to as ice creams.

History

Most early ice cream recipes would be called frozen custard today, because of naming conventions. The modern term of frozen custard is primarily used as a method of brand differentiation and to comply with FDA requirements. One claim traces the invention of the term frozen custard to Coney Island in 1919. However, recipes for custard based ices existed before then, especially in France. In some areas of the United States, "frozen custard" or "custard" has become a synonym for soft serve.

Home Creation

The vast majority of ice cream desserts created at home are frozen custards. Despite commercialization of a limited number of flavours, home cooks can make frozen custard into innumerable flavours. Extra egg-yolk and a brief heating is used in order to make a custard base for the ice. This replaces commercially used ice cream additives such as gluten in order to create the same consistency. It is possible that the heating of the cream containing milk was initially done as a form of pasteurization, with the side effect of creating a custard. At home, frozen custard can be made with a standard ice cream maker or with the freeze and scrape method.

Commercial Creation

Using a process called overrun, air is blended into the mixture of ingredients until its volume increases by approximately 20%. By comparison, ice cream may have an overrun as large as 100%, meaning half of the final product is composed of air. The high percentage of butterfat and egg yolk gives frozen custard a thick, creamy texture and a smoother consistency than ice cream. Frozen custard can be served at –8°C (18°F), warmer than the –12°C (10°F) at which ice cream is served, in order to make a soft serve product.

Another difference between commercially produced frozen custard and commercial ice cream is the way the custard is frozen. The mix enters a refrigerated tube and, as it freezes, blades scrape the product cream off the barrel walls. The now frozen custard is discharged directly into containers from which it can be served. The speed with which the product leaves the barrel minimizes the amount of air in the product but more importantly ensures that the ice crystals formed are very small.

Frozen custard is usually prepared fresh at the place of sale, rather than stored; however, it is occasionally available in supermarkets or by mail order. Generally, modern frozen custard stands provide only three different flavours per day: vanilla, chocolate, and a unique "flavour of the day." The older vintage custard stands tend to have a dozen or so standard flavours. Some locations try to emulate the Coney Island feel (sometimes cited as the birthplace of the term frozen custard) by serving other diner foods along with their premium dessert.

Semifreddo, a Class of Semi-frozen Desserts

Semifreddo (Italian: "half cold") is a class of semi-frozen desserts, typically ice-cream cakes, semi-frozen custards, and certain fruit tarts. It has the texture of frozen mousse because it is usually produced by uniting two equal parts of ice cream and whipped cream. Such a dessert's Spanish counterpart is called semifrío. In Italian cuisine, the Semifreddo is commonly made with gelato as a primary ingredient. It is typical of the Italian region of Emilia-Romagna.

Sorbet

Sorbet is a frozen dessert made from sweetened water flavoured with fruit (typically juice or puree), wine, and/or liqueur. The origin of sorbet is variously explained as either a Roman invention, or a Middle Eastern drink *charbet*, made of sweetened fruit juice and water. The term *sherbet* or *charbet* is derived from Turkish: *þerbat*/°erbet, "sorbet", from the Persian *sharbat*. Sorbet is sometimes served between courses as a way to cleanse the palate before the main course.

Classification and Description

Sorbet is often confused with Italian ice and often taken to be the same as sherbet.

Sorbets/sherbets may also contain alcohol, which lowers the cold temperature, resulting in softer texture. In the UK, *sherbet* refers to a fizzy powder, and only the term *sorbet* would be used.

Whereas ice cream is based on dairy products with air copiously whipped in, sorbet has neither, which makes for a dense and extremely flavorful product. Sorbet is served as a non-fat or low-fat alternative to ice cream.

In Italy a similar though crunchier textured dish called granita is made. As the liquid in granita freezes it forms noticeably large-size crystals, which are left unstirred. Granita is also often sharded with a fork to give an even crunchier texture when served.

Agraz is a type of sorbet, usually associated with the Maghreb and north Africa. It is made from almonds, verjuice, and sugar. It has a strongly acidic flavour, because of the verjuice. (*Larousse Gastronomique*)

Early History and Folklore

One account says that Marco Polo brought a recipe for a sorbet-like dessert on his way back to Italy from China in the late 13th century, as written in an account of his journey, *The Travels of Marco Polo*.

Other folklore holds that Nero, the Roman Emperor, invented sorbet during the first century A.D. when he had runners along the Appian way pass buckets of snow hand over hand from the mountains to his banquet hall where it was then mixed with honey and wine.

Frozen desserts are believed to have been brought to France in 1533 by Catherine de' Medici when she left Italy to marry the Duke of Orleans, who later became Henry II of France. By the end of the 17th century, sorbet was served in the streets of Paris, and spread to England and the rest of Europe.

Distinction from Sherbet

American Terminology

In the United States, sorbet and sherbet (although it's spelled 'sherbet' it is widely pronounced 'sherbert') are distinctly different products. For Americans, sherbet is the more widely-known term and typically designates a fruity flavoured frozen dairy product with a milkfat content between 1 and 2%. Sorbet, on the other hand, is considered by Americans to be a fruity frozen product with no dairy content, similar to Italian ice.

Sherbet in the United States must include dairy ingredients such as milk or cream to reach a milkfat content between 1% and 2%. Products with higher milkfat content are defined as ice cream; products with lower milkfat content are defined as water ice. The use of the term "sorbet" is unregulated and is most commonly used with non-dairy, fruit juice "italian ice" products. Although the American legal definitions indicate that the terms "sorbet" and sherbet are interchangeable, actual usage by Americans and the manufacturers of these products bears a clear distinction. A similar situation occurs in the legal definitions by differing international state governments on what is considered beer.

English/French Labeling

On sherbet packages which have both English and French labels, sherbet is translated to *sorbet laitier* which directly translates into English as *dairy sorbet*, differentiating the milk-containing sherbet from milk-less sorbet.

Kulfi

Kulfi or Qulfi is a popular frozen milk-based dessert from India and Pakistan that is often described as "traditional Indian ice-cream". It is popular throughout neighboring countries in South Asia, Burma (Myanmar), and even the Middle East. It has similarities to ice cream (as popularly understood) in appearance and taste, but is denser and creamier.

It comes in various flavours, including cream (malai), raspberry, rose, mango, cardamom (*elaichi*), saffron (*kesar* or *zafran*), the more traditional flavours, as well as newer variations like apple, orange, strawberry, peanut, and avocado. Unlike Western ice creams, kulfi is not whipped, resulting in a solid, dense frozen dessert similar to traditional custard based ice-cream. Thus, it is sometimes considered a distinct category of frozen dairy-based dessert. Due to its density, kulfi takes a longer time to melt than Western ice-cream.

History

Just like any other culture exposed to snow and ice, some people living in the Indian subcontinent, especially those living high in the Himalayas, would have stumbled upon the technique of freezing various sweetened liquids, thus turning them into frozen desserts. These privileges were limited to royalty and upper levels of aristocracy in India until modern day refrigeration technology reached South Asia.

Preparation

Kulfi was traditionally prepared by evaporating sweetened and flavoured milk by slow cooking, with almost continuous stirring to keep milk from sticking to the bottom of the vessel where it might burn, till its volume was reduced by a half, thus thickening it, increasing its fat, protein and lactose density. It has a distinctive taste due to caramelization of lactose and sugar during the lengthy cooking process. The semi-condensed mix is then frozen in tight sealed molds (often kulhars with their mouths sealed) that are then submerged in ice mixed with salt to speed up the freezing process. The ice/salt mix, along with its submerged kulfi molds, is placed in earthen pots or matkas that provide insulation from the external heat and slow down the melting of ice. Kulfi prepared in this manner is hence called 'Matka Kulfi'. Kulfi, thus prepared by slow freezing, also renders a unique smooth mouth feel that is devoid of water crystallization.

More recently Kulfi is prepared from evaporated milk, sweetened condensed milk and heavy (double) cream. Then sugar is added and the mixture is further boiled and cornstarch-water paste is added. This paste thickens the mixture, although it is boiled for an additional few minutes. Then flavourings, dried fruits, cardamom, etc. are added. The mixture is then cooled, put in moulds and frozen. If frozen in individual-portion custard bowls for service with a spoon, bowls are removed from the freezer 10–15 minutes before serving to allow for melting at the edges.

It is garnished with ground cardamom, saffron, or pistachio nuts. Kulfi is also served with faloodeh (vermicelli noodles made from starch). In some places, people make it at home and make their own flavours.

Traditionally in India and Pakistan, kulfi is sold by street vendors called *kulfiwallahs* who keep the kulfi frozen by placing the moulds inside a large earthenware pot called a "matka", filled with ice and salt. It is served on a leaf or frozen onto a stick. It can be garnished with pistachios, cardamom and similar. Often it is served as Falooda Kulfi which is kulfi with rice noodles, rose syrup and other ingredients. Popular flavours include pistachio, mango, vanilla, and rose.

Spaghettieis

Spaghettieis is a German ice cream specialty that looks like a plate of spaghetti. It was invented by Dario Fontanella in the late 1960s in Mannheim, Germany. In the dish, an often light or white colored ice cream is pressed through a modified Spätzle press or potato ricer to make it look like spaghetti. It is then placed over whipped cream and topped with strawberry sauce (to simulate tomato sauce) and either coconut flakes, grated almonds, or white chocolate shavings to represent the parmesan cheese. Although, as of yet, it is not well known to most people outside Europe, it can be found at some Gelaterias and specialty ice cream shops, at special events and at some hotels and restaurants around the world. Spaghettieis is widely recognized across Germany, and it costs around €3,50 a dish.

Granita

Granita (in Italian also granita siciliana) is a semi-frozen dessert made from sugar, water and various flavourings. Originally from Sicily, although available all over Italy (but granita in Sicily is somewhat different from the rest of Italy), it is related to sorbet and italian ice. However, in most of Sicily, it has a coarser, more crystalline texture. Food writer Jeffrey Steingarten says that "the desired texture seems to vary from city to city" on the island; on the west coast and in Palermo, it is at its chunkiest, and in the east it is nearly as smooth as sorbet. This is largely the result of different freezing techniques: the smoother types are produced in a gelato machine, while the coarser varieties are frozen with only occasional agitation, then scraped or shaved to produce separated crystals.

Ingredients

Common and traditional flavoring ingredients include lemon juice, mandarin oranges, jasmine, coffee, almonds, mint, and when in season wild strawberries and black mulberries. Chocolate granitas have a tradition in the city of Catania and, according to Steingarten, nowhere else in Sicily. The nuances of the Sicilian ingredients are important to the flavour of the finished granita: Sicilian lemons are a less acidic, more floral variety similar to Meyer lemons, while the almonds used contain some number of bitter almonds, crucial to the signature almond flavour.

Serving Conventions

Granita with coffee is very common in the city of Messina, while granita with almonds is popular in the city of Catania. Granita in combination with a yeast pastry called *brioche* is a common breakfast in summer time. (The Sicilian brioche is generally flatter and wider than the French version.)

Granita is often found served as a slush-type drink rather than a dessert, in a paper or plastic cup with a plastic lid and a straw (often a spoon straw).

Frozen Yogurt

Frozen yogurt (also known as frozen yoghurt or froz yog or by the tradenames FroYo and Frogurt) is a frozen dessert made from, or containing yogurt or other dairy products. It is slightly more tart than ice cream, as well as lower in fat (due to the use of milk instead of cream). It differs from ice milk (more recently termed low-fat or light ice cream), which does not include yogurt as an ingredient.

History

Frozen yogurt was introduced in New England in the 1970s as a soft serve dessert by H. P. Hood under the name *Frogurt*. In 1978,

Brigham's, a Boston-based ice cream, candy and sandwich chain, developed and introduced the first packaged frozen yogurt under the name *Humphreez Yogart.*

It was originally intended as a healthier alternative to ice cream, but consumers complained about the tart taste. Manufacturers began production of a recipe that tasted sweeter, and frozen yogurt took off in the 1980s, reaching sales of $25 million in 1986. In the early 1990s, frozen yogurt was 10% of the frozen dessert market. The first frozen yogurt shop in the UK, Lick, based in Brighton opened in March 2008. Later that year, frozen yogurt shops began opening in London and across the country.

Production

Frozen yogurt usually consists of milk solids, milk fat, yogurt culture, sweetener, corn syrup, colouring, and flavoring.

Milk fat comprises about 0.5-6% of the yogurt. Added in quantities inversely proportional to the amount of milk solids, the milk fat lends richness to the yogurt. Milk solids account for 8-14% of the yogurt's volume, providing lactose for sweetness and proteins for smoothness and increased resistance to melting. Cane or beet sugar provides 15-17% of the yogurt's ingredients. In addition to adding sweetness, the sugar increases the volume of solid ingredients in the yogurt, improving body and texture. Animal gelatins and vegetable additives (guar gum, carrageenan, etc.) stabilize the yogurt, reducing crystallization and increasing the temperature at which the yogurt will melt. This stabilization ensures that the frozen yogurt maintains a smooth consistency regardless of handling or temperature change.

Frozen yogurt can be made in an ice cream machine; however, major companies often use assembly lines specifically dedicated to frozen yogurt production. The milk products and gelatin are combined and homogenized. They are then cooled to 0 degrees celsius. At this temperature, the yogurt culture is added and the mixture is warmed to 4 degrees Celsius.

Once it has reached the desired temperature and viscosity, the yogurt is allowed to sit in aging tanks for up to four hours. Sweeteners, colorings and flavourings are then mixed in, and the yogurt mixture is cooled to -6 to -2 degrees Celsius. To create extra volume and smooth consistency, air is incorporated into the yogurt as the mixture is agitated. When a sufficient amount of air has been incorporated into the product, the yogurt is rapidly frozen to prevent the formation of large ice crystals, and stored in a cold place to be shipped.

Uses

Frozen yogurt has come to be used much like ice cream, and is served in a wide variety of flavours and styles. Many companies allow customers the option of adding various toppings, or ordering their frozen yogurt in cups or in cones. Certain sellers even offer sugar-free varieties. Frozen yogurt made by chains such as Yogurt Fusion, Yogurt Story, Yoforia, Pinkberry, Red Mango, and Yogen Früz is tarter and closer to the original recipe, whereas other companies like TCBY and I Can't Believe It's Yogurt focus on making their frozen yogurt taste like ice cream.

Franchising

It is expected to be a multi-billion dollar industry within the next few years, creating a prime opportunity for franchise expansion. "Which franchise categories holds the most promise for 2009. Frozen yogurt is also turning out to be a simply irresistible opportunity", Entrepreneur Magazine. In 1999, Entrepreneur Magazine rated Yogen Fruz the number 1 franchise in the world that year, ahead of perennial winners such as Subway and McDonald's. Popular national frozen yogurt franchises include Yogen Früz, Pinkberry, Menchie's, Red Mango, TCBY, Cherry on Top and Tutti Frutti.

Ghee

Ghee is a class of clarified butter that originated in South Asia, and is commonly used in South Asian (Indian, Bangladeshi, Nepali and Pakistani), North African (Egyptian and Berber) and Horn African cuisine (Somali, Ethiopian and Eritrean).

Preparation

Ghee, also known as *clarified butter* in anglophone countries, is made by simmering unsalted butter in a cooking vessel until all water has boiled off, the milk solids (or protein) have settled to the bottom, and a scum has floated on top. After removing the scum, the cooked and clarified butter is then spooned off or tipped out carefully to avoid disturbing the milk solids on the bottom of the pan. Ghee can be stored for extended periods without refrigeration, provided that it is kept in an airtight container to prevent oxidation and remains moisture-free. The texture, colour, or taste of ghee depends on the source of the milk from which the butter was made and the extent of boiling and simmering.

Religious Use

Ghee made from cow's milk has a sacred role in Vedic and modern Hindu libation and anointment rituals. There is also a hymn to ghee.

Ghee is also burnt in the Hindu religious ritual of *Ârati* (Aarti) and is the principal fuel used for the Hindu votive lamp known as the *diyâ* or *dîpa* (deep). It is used in marriages and funerals, and for bathing *mûrtis* (divine idols) during worship.

In other religious observances, such as the prayers to *Œiva* (Shiva) on *Mahâ-úivarâtrî* (Maha Shivaratri), ghee is served in *Pañcâm[ta* (Panchamruta) along with four other sacred substances: sugar, honey, milk, and *dahî* (yogurt). According to the *Mahâbhârata*, ghee is the very root of sacrifice by Bhîcma. Also, it is used generously in *homam* or *yajña* since it is considered as food for the Devas.

Usage in Food

Ghee is widely used in Indian cuisine and Pakistani cuisine. It, however, is mentioned in the Epic of Gilgamesh, and is probably Akkadian in origin. In many parts of India and Pakistan, especially in Punjab, Haryana, Gujarat, Maharashtra, Bengal, Orissa and many other states, rice is traditionally prepared or served with ghee (including biryani). In the Bharuch district of Gujarat, Ghee is served with kichri, usually an evening meal of yellow rice with curry, a sauce made from yoghurt, cumin seeds, curry leaves, ghee, cornflour, tumeric, garlic and salt. Ghee is also an ingredient as well as used in the preparation of *kadhi* and used in Indian and Pakistani sweets such as Mysore pak, and different varieties of halva and laddu. Punjabi cuisine prepared in restaurants uses large amounts of ghee. Naan and roti are sometimes brushed with ghee, either during preparation or while serving.

Ghee is an ideal fat for deep frying because its smoke point (where its molecules begin to break down) is 250°C (485°F), which is well above typical cooking temperatures of around 200°C (400°F) and above that of most vegetable oils.

Nutrition

Like any clarified butter, ghee is composed almost entirely of fat; the nutrition facts label found on bottled cow's ghee produced in the USA indicates eight mg. of cholesterol per teaspoon.

Ghee has been shown to slightly, but not significantly, reduce serum cholesterol in one rodent study. Studies in Wistar rats have revealed one mechanism by which ghee reduces plasma LDL cholesterol. This action is mediated by an increased secretion of biliary lipids.

Indian restaurants and some households may use hydrogenated vegetable oil (also known as *vanaspati*, *dalda*, or "vegetable ghee") in place of ghee due to its lower cost. This "vegetable ghee" may contain trans fat. Trans fats are increasingly linked to serious chronic health conditions. The term *shuddh ghee*, however, is not officially enforced in many regions, so partially hydrogenated oils are marketed as pure ghee in some areas. Where this is illegal in India, law enforcement often cracks down on the sale of fake ghee. Ghee is also sometimes called *desi* (country-made) ghee or *asli* (genuine) ghee to distinguish it from "vegetable ghee".

Outside South Asia

Several cultures make ghee outside of South Asia. Egyptians make a product called *samna baladi*, literally meaning "local ghee"; i.e., Egyptian ghee virtually identical to ghee in terms of process and end result. In Ethiopia, [[*niter kibbeh*]] is made and used in much the same way as ghee, but with spices added during the process that result in a distinctive taste. Moroccans (especially Berbers) take this one step further, aging spiced ghee in the ground for months or even years, resulting in a product called *smen*. In northeastern Brazil, an unrefrigerated butter very similar to ghee, called *manteiga-de-garrafa* (butter-in-a-bottle) or *manteiga-da-terra* (butter of the land), is common. In Europe, it is also widely used. For example, *Wiener Schnitzel* is traditionally fried in a version of ghee called *Butterschmalz.*

Whey

Whey or milk plasma is the liquid remaining after milk has been curdled and strained. It is a by-product of the manufacture of cheese or casein and has several commercial uses. *Sweet whey* is manufactured during the making of rennet types of hard cheese like Cheddar or Swiss. *Acid whey* (also known as "sour whey") is obtained during the making of acid types of cheese such as cottage cheese.

Production

Whey is a co-product of cheese production. It is one of the components that separate from milk after curdling, when rennet or an edible acidic substance is added.

Uses

Whey is used to produce ricotta, brown cheeses, Messmör/Prim, and many other products for human consumption. It is also an additive in many processed foods, including breads, crackers, and commercial pastry, and in animal feed. Whey proteins consist primarily of α-lactalbumin and β-lactoglobulin. Depending on the method of manufacture, whey may also contain glycomacropeptides (GMP).

Whey protein (derived from whey) is often sold as a nutritional supplement. Such supplements are especially popular in the sport of bodybuilding. In Switzerland, where cheese production is an important industry, whey is used as the basis for a carbonated soft drink called Rivella. In Iceland, MS manufactures and sells liquid whey as *Mysa* in 1-liter cartons (energy 78 kJ or 18 kcal, calcium 121 mg, protein 0.4 g, carbohydrates 4.2 g, sodium 55 mg).

Throughout history, whey was a popular drink in inns and coffee houses. When Joseph Priestley was at college at Daventry Academy 1752–1755, he records that, during the morning of Wednesday 22 May 1754, he "went with a large company to drink whey." This was probably 'sack whey' or 'wine whey.' A contemporary recipe for 'wine whey' instructs: "Put a pint of skimmed milk, and half a pint of white wine into a bason, let it stand a few minutes, then pour over it a pint of boiling water, let it stand a little, and the curd will gather in a lump, and settle to the bottom, then pour your whey into a China bowl, and put in a lump of sugar, a sprig of balm, or a slice of lemon."

An alternative recipe is for 'Cream of Tartar Whey': "Put a pint of blue milk over the fire, when it begins to boil, put in two tea spoonfuls of cream of tartar, then take it off the fire, and let it stand till the curd settles to the bottom of the pan, then put it into a bason to cool, and drink it milk warm." Whey was also used in central Spain to enrich bakery products. In some traditions, it was used instead of water to produce bread dough.

Whey Cream and Butter

Cream can be skimmed from whey. Whey cream is more salty, tangy, and "cheesy" than ("sweet") cream skimmed from milk, and can be used to make whey butter. Whey cream and butter are suitable for making butter-flavoured food, as they have a stronger flavour of their own. They are also cheaper than sweet cream and butter.

Health

Because whey contains lactose, it should be avoided by those who are lactose intolerant. Dried whey, a very common food additive, contains more than 70% lactose. When used as a food additive, whey can contribute to quantities of lactose far above the level of tolerance of most lactose-intolerant individuals.

Liquid whey contains lactose, vitamins, protein, and minerals, along with traces of fat. In 2005, researchers at Lund University in Sweden discovered that whey appears to stimulate insulin release, in type 2 diabetics. Writing in the *American Journal of Clinical Nutrition*, they also discovered that whey supplements can help regulate and reduce spikes in blood sugar levels among people with type 2 diabetes by increasing insulin secretion.

Protein

Whey protein is the name of globular proteins that can be isolated from whey. It is typically a mixture of globinstagers beta-lactoglobulin (~65%), alpha-lactalbumin (~25%), and serum albumin (~8%), which are soluble in their native culture forms, independent of pH.

Casein and Caseinates

Casein (from Latin *caseus*, "cheese") is the name for a family of related Phosphoprotein proteins. These proteins are commonly found in mammalian milk, making up 80% of the proteins in cow milk and between 20% to 45% of the proteins in human milk. Casein has a wide variety of uses, from being a major component of cheese, to use as a food additive, to a binder for safety matches. As a food source, casein supplies essential amino acids as well as some carbohydrates and the inorganic elements calcium and phosphorus.

Description

Casein consists of a fairly high number of proline peptides, which do not interact. There are also no disulfide bridges. As a result, it has relatively little tertiary structure. Because of this, it cannot denature. It is relatively hydrophobic, making it poorly soluble in water. It is found in milk as a suspension of particles called casein micelles which show some resemblance with surfactant-type micellae in a sense that the hydrophilic parts reside at the surface.

The caseins in the micelles are held together by calcium ions and hydrophobic interactions. There are several models that account for the special conformation of casein in the micelles. One of them proposes that the micellar nucleus is formed by several submicelles, the periphery consisting of microvellosities of ê-casein.

Another model suggests that the nucleus is formed by casein-interlinked fibrils (Holt, 1992). Finally, the most recent model (Horne, 1998) proposes a double link among the caseins for gelling to take place. All 3 models consider micelles as colloidal particles formed by casein aggregates wrapped up in soluble ê-casein molecules.

The isoelectric point of casein is 4.6. Since milk's pH is 6.6, casein has a negative charge in milk. The purified protein is water insoluble.

While it is also insoluble in neutral salt solutions, it is readily dispersible in dilute alkalis and in salt solutions such as sodium oxalate and sodium acetate.

The enzyme trypsin can hydrolyze off a phosphate-containing peptone. It is used to form a type of organic adhesive.

Uses

Paint

Casein paint, is a fast-drying, water-soluble medium used by artists. Casein paint has been used since ancient Egyptian times as a form of tempera paint, and was widely used by commercial illustrators as the material of choice until the late 1960s when, with the advent of acrylic paint, casein became less popular.

Glue

Casein-based glues were popular for woodworking, including for aircraft as late as the de Havilland Mosquito.

Cheese making

Cheese consists of proteins and fat from milk, usually the milk of cows, buffalo, goats, or sheep. It is produced by coagulation of casein. Typically, the milk is acidified and then coagulated by the addition of rennet, a proteolytic enzyme typically obtained from the stomachs of calves. The solids are separated and pressed into final form.

Unlike many proteins, casein is not coagulated by heat. During the process of clotting, milk-clotting proteases act on the soluble portion of the caseins, K-Casein, thus originating an unstable micellar state that results in clot formation.

When coagulated with chymosin, casein is sometimes called paracasein. Chymosin (EC 3.4.23.4) is an aspartic protease that specifically hydrolyzes the peptide bond in Phe105-Met106 of ê-casein and is considered to be the most efficient protease for the cheese-making industry (Rao et al., 1998).

British terminology, on the other hand, uses the term caseinogen for the uncoagulated protein and casein for the coagulated protein. As it exists in milk, it is a salt of calcium.

Plastics and Fiber

Some of the earliest plastics were based on casein. In particular Galalith was well-known for use in buttons. Fiber can be made from extruded casein.

Protein Supplements

An attractive property of the casein molecule is its ability to form a gel or clot in the stomach. The ability to form this clot makes it very efficient in nutrient supply.

The clot is able to provide a sustained slow release of amino acids into the blood stream, sometimes lasting for several hours. This provides better nitrogen retention and utilization by the body. Plasma immunoreactive IGF-1 concentration in rats given a casein diet was higher than that in rats given a soya-bean-protein or protein-free diet.

Medical and Dental Uses

Casein-derived compounds are used in tooth remineralization products to stabilize amorphous calcium phosphate (ACP) and release the ACP onto tooth surfaces where it can facilitate remineralization.

Controversies

Autism

Casein has been documented to break down to produce the peptide casomorphin, an opioid that appears to act primarily as a histamine releaser. Some research indicates that this casomorphine aggravates the symptoms of autism. A 2006 review of seven studies indicated that although benefits were seen in all studies from the introduction of elimination diets (e.g., casein or gluten free) in the treatment of autism spectrum disorders, none of the studies were performed in a manner to create an unbiased scientific opinion.

Preliminary data from the first and only double-blind randomized control trial of a gluten- and casein-free diet "indicated no statistically significant findings even though several parents reported improvement in their children." Although research has shown high rates of use of complementary and alternative therapies (CAM) for children with autism including gluten and/or casein exclusion diets, the evidence for efficacy of these diets is currently unsubstantiated.

A1/A2 Beta Caseins

Four casein proteins make up about 80% of the protein in cow's milk. One of the major caseins is beta-casein, of which there are several types, but "A1" and "A2" are the most common. According to the Australian Food Standards Agency. Certain breeds of cows, such as Friesians, produce mostly A1 milk, whereas other breeds, such as Guernseys, as well as sheep and goats, produce mostly A2 milk.

Food Standards Australia New Zealand also reports that despite some hypotheses that consumption of A2 milk "provides levels of protection to consumers from autism in children as well as schizophrenia, diabetes and heart disease," the scientific evidence for such claims is "very limited". Additionally, the European Food Safety Authority carried out a literature review in 2009 concluding that "a cause and effect relationship is not established between the dietary intake of BCM7, related peptides or their possible protein precursors and non-communicable diseases". Studies supporting these claims have had significant flaws, and the data are inadequate to guide autism treatment recommendations.

T. Colin Campbell's *The China Study* (2006), a univariate correlative published book, suggests a correlation between powdered, isolated casein administered to rats and the promotion of cancer cell growth when exposed to carcinogens. A 2001 study suggests that another milk protein, whey protein, but not casein, may play a protective role against colon tumors in rats. According to a study from the Australian Dairy Council, casein has antimutagenic effects.

Casein-free Diet

Casein has a molecular structure that is quite similar to that of gluten. Thus, some gluten-free diets are combined with casein-free diets and referred to as a gluten-free, casein-free diet. Casein is often listed as sodium caseinate, calcium caseinate or milk protein. These are often found in energy bars, drinks, and packaged goods. A small fraction of the population is allergic to casein.

Altering the Effects of Polyphenols

A study of Charité Hospital in Berlin by Lorenzo et al., published in The European Heart Journal, showed that adding milk to tea causes the casein to bind to the molecules in tea that cause the arteries to relax, especially a catechin molecule called EGCG, although a more recent study by Reddy et al. (2005) suggests that the addition of milk to tea does not alter the antioxidant activity in vivo and the cardiovascular effect remains controversial. A study published in the journal *Free Radical Biology and Medicine* indicates that casein reduced the peak plasma levels of beneficial polyphenols after the consumption of blueberries.

Casein Products

Casein is the principal protein found in cow.s milk from which it has been extracted commercially for most of the 20th century. It is responsible for the white, opaque appearance of milk in which it is combined with calcium and phosphorus as clusters of casein molecules, called .micelles.

The major uses of casein until the 1960s were in technical, non-food applications such as adhesives for wood, in paper coating, leather finishing and in synthetic fibres, as well as plastics for buttons, buckles *etc.* During the past 30 years, however, the principal use of casein products has been as an ingredient in foods to enhance their physical (so-called .functional.) properties, such as whipping and foaming, water binding and thickening, emulsification and texture, and to improve their nutrition.

In New Zealand, casein is precipitated from the skim milk that is produced after centrifugal separation of whole milk. The skim milk may be acidified to produce acid casein or treated with an enzyme, resulting in the so-called rennet casein. The precipitated casein curd is separated from the whey, washed and dried. Water-soluble derivatives of acid caseins, produced by reaction with alkalis, are called caseinates.

The amount of casein in cow.s whole milk varies according to the breed of cow and stage of lactation, but is generally in the range 24-29 g L-1. Casein contains 0.7-0.9% phosphorus, covalently bound to the protein by a serine ester linkage. Casein is consequently known as a phospho-protein. All the amino acids that are essential to man are present in casein in high proportions, with the possible exception of cysteine. Thus, casein may be considered as a highly nutritious protein.

Casein exists in milk in complex groups of molecules (sometimes referred to as calcium phospho-caseinate) that are called .micelles.. The micelles consist of casein molecules, calcium, inorganic phosphate and citrate ions, and have a typical molecular weight of several hundred million. In terms of physical chemistry, the casein micelles may be considered to exist in milk as a very stable colloidal dispersion.

The caseins, as proteins, are made up of many hundreds of individual amino acids, each of which may have a positive or a negative charge, depending on the pH of the [milk] system. At some pH value, all the positive charges and all the negative charges on the [casein] protein will be in balance, so that the net charge on the protein will be zero. That pH value is known as the isoelectric point (IEP) of the protein and is generally the pH at which the protein is least soluble. For casein, the IEP is approximately 4.6 and it is the pH value at which acid casein is precipitated. In milk, which has a pH of about 6.6, the casein micelles have a net negative charge and are quite stable.

Although casein has been shown to consist of several individual casein components, referred to as αs1-, αs2-, α- and β-casein, each having slightly different properties (which are caused by small variations in their amino acid content).

Extraction of Casein from Milk

Separation

Whole cow.s milk (with a typical fat content of 4.65%) is first separated by means of centrifuges that produce cream (for the manufacture of butter or other milkfat products) and skim milk. Skim milk can thus be considered as the raw material from which casein products are made.

Precipitation

Precipitation by means of acidification can be considered in terms of simple chemistry as follows, R being the casein protein:

H2N-R-COO- + H+ → +H3N-R-COOcasein micelle acid casein (pH = 6.6) (pH = 4.6)

Colloidal Dispersion Insoluble Particles

In the case of enzyme coagulation of casein, there is no change in the pH of the milk. Instead, the addition of a specific enzyme, *chymosin*, which is found in the stomach of newborn calves, specifically cleaves a highly charged portion from the ê-casein, called glycomacropeptide. That action causes the remainder of the ê-casein (now called *para*-ê-casein) to lose its considerable power in stabilising the micelles in milk, and the result is the formation of a three-dimensional gel network or clot of the casein in the presence of calciumions. This reaction is essential in the manufacture of virtually all cheese types and in the production of rennet casein.

Wet-processing Operations

When the casein has been precipitated, the mixture is heated (a process known in the dairy industry as .cooking.). Heating of the precipitated casein causes the particles to shrink and expel moisture (whey) (rather like a sponge), and also to agglomerate together to form clumps of curd. The curd is then separated from the whey and washed several times with water in vats prior to mechanical dewatering by pressing or centrifuging.

Drying and Dry Processing of Casein

The dewatered curd, with a moisture content of about 55%, is dried by means of hot air using either fluidised bed driers with multiple decks or pneumatic-conveying ring driers to produce a dry casein having a moisture content of 10-12%. The warm, unmilled casein is then subjected to several dry processing steps, including cooling (usually by air conveying), .tempering. or conditioning to ensure that moisture is distributed evenly between large and small particles, milling, sifting (to produce coarse, medium and fine mesh particles), blending (to ensure uniformity) and bagging. The 25 kg bags of casein are placed on pallets and stored ready for shipping.

Types of Casein

As indicated above, two basic types of casein - acid and rennet - are produced in New Zealand. They are named in accordance with the coagulating agent employed. Three types of acid casein are made commercially: lactic, hydrochloric and sulphuric acid caseins. In New Zealand, lactic acid casein has been the most common casein product, although larger quantities of sulphuric acid casein have been produced in recent years.

In Australia and Europe, the most common precipitant for acid casein is hydrochloric acid, which is a by-product of the chemical industry and hence is relatively cheap. In New Zealand, however, which has a very small chemical industry, hydrochloric acid is relatively expensive. On the other hand, sulphuric acid is relatively cheap, being produced in comparatively large quantities by the fertiliser industry for use in the manufacture of superphosphate. Consequently, virtually all mineral acid casein made in New Zealand is precipitated using sulphuric acid. The properties of the different types of acid casein are very similar and, for most applications, the acid caseins can be used interchangeably.

For the manufacture of rennet casein, several different coagulants are now available. These include chymosin (previously known as .rennet. or .rennet extract.), the milk-clotting enzyme extracted from the stomachs of young calves, and a number of so-called microbial rennets, which are enzymes that have been produced by means of microbial fermentation techniques. The caseins produced using any of these enzyme preparations are all known as rennet casein, and all have similar properties. However, their properties are noticeably different from those of acid casein.

Acid Casein Manufacture

Lactic Acid Casein

For the manufacture of lactic acid casein, skim milk (pH 6.6) is first pasteurised (72°C for 15 s). It is then cooled to setting temperature (22-26°C) and inoculated with several strains of lactic acid-producing bacteria, known as .starters. (*e.g. Lactococcus lactis* sub-species *cremoris*, 0.1-0.2% of milk volume). The milk is incubated, without agitation, in large silos (each with a capacity of up to 250 000 L) for a period of 14-16 h. During this period, some of the lactose in the milk is converted to lactic acid by the starter and the pH is reduced to about 4.6, causing coagulation of the casein (and the milk). This takes the form of a soft gel and is generally referred to in the industry as .coagulum. or .coag.

The fermentation of lactose to lactic acid in the manufacture of lactic acid casein is not as simple. however, and a number of other compounds are produced as well, *e.g.* diacetly (CH3COCOCH3), acetoin (CH3CHOHCOCH3) and benzoin (C6H5COCHOHC6H5).

These are present in relatively small amounts and do not generally present any serious problems.

After the pH of the milk has reached 4.6-4.7, the coagulum is .cooked. (*i.e.* heated), usually by means of a combination of indirect heating (through a heat exchanger) and steam injection, to a temperature of 50-55oC. After a brief period of residence in a cooking line and acidulation.

Mineral Acid Casein

For the precipitation of mineral acid casein, pasteurised skim milk at a pH of 6.6 is mixed thoroughly with dilute (0.25 mol L-1) acid at a temperature of about 20°C to a pH of approximately 4.6. In this case, because of the very vigorous agitation and the short mixing time, the casein is precipitated as very fine, individual particles in a liquid serum (whey), unlike the gel/coagulum formed in lactic acid casein manufacture. The acidified milk mixture is then cooked and processed further in a manner similar to that described for the production of lactic acid casein.

Rennet Casein Manufacture

When rennet casein is made, the skim milk is not acidified. Hence, the pH remains at 6.6 throughout the manufacturing process. Following pasteurisation of the skim milk, it is cooled to a setting temperature of about 29°C, and calf rennet or microbial rennet is added (ratio by volume: 1 of rennet to about 7000 of skim milk) and mixed in thoroughly. During the first stage of renneting, the enzyme specifically cleaves one of the bonds in ê-casein, releasing part of the protein chain that is commonly referred to as glycomacropeptide. This action destabilises the casein micelles and, in the second stage of the reaction, they form a three-dimensional clot with some of the calcium ions that are present in the milk. The renneting process usually takes place in a period of 20-40 min under the conditions (pH and temperature) described above. The clotted milk may then be cooked and the casein processed in a manner similar to that described for lactic acid casein.

Yield

The yield of commercial casein is about 3 kg/100 kg skim milk.

Manufacture of Caseinates

Caseinates may be produced by reaction under aqueous conditions of acid casein curd or dry acid casein with any one of several different dilute alkalis.

The resulting homogeneous solution may be spray dried to produce a caseinate powder having a moisture content of 3-6%, depending on the manufacturing conditions and customer requirements.

Spray-dried Sodium Caseinate

The most common alkali used in the manufacture of spray dried sodium caseinate is sodium hydroxide. It is mixed (as an aqueous solution with a typical concentration of 2.5 M) with a slurry of the casein curd or powder in water. The usual amount of sodium hydroxide needed is about 2% (w/w) of the casein solids.

The casein curd is milled using one or more colloid mills to reduce the size of the individual particles so that they will dissolve rapidly, and is then mixed with the alkali using high shear. In producing solutions of sodium caseinate for spray drying, it is important to achieve the maximum possible concentration of solids for economic reasons, as the more water there is to evaporate, the higher is the energy cost. Concentrated solutions of sodium caseinate (> 15% solids), however, are very viscous, and require powerful agitators and pumps for mixing and fluid transfer. The use of high temperatures (60-95°C) during the later dissolving stages is of practical benefit, as the viscosity of sodium caseinate solutions decreases with temperature. However, there is still a delicate balance between what is mechanically achievable and what is economically practicable during the commercial production of sodium caseinate.

Other Caseinates

The manufacture of potassium and ammonium caseinates is very similar to that of sodium caseinate, although, in the case of ammonium caseinate, a lot of the ammonia is evaporated from the solution during the drying process. A solution of sodium caseinate, like those of potassium and ammonium caseinates, has a straw-like colour and is completely different in appearance from milk. Solutions of calcium caseinate, on the other hand, are very white and opaque - even whiter than milk, and they are less viscous than solutions of the other caseinates. Calcium caseinate solutions are produced by adding a slurry of lime (calcium hydroxide) in water to a casein curd-water mixture and reacting the combined slurries at a relatively low temperature (< 45°C) until the neutralisation reaction is completed. Use of higher temperatures before neutralisation is completed is likely to result in precipitation or coagulation of the partly reacted calcium caseinate, with probable dumping of the contents of the reaction vessel. All caseinate powders have a white appearance.

Composition of Casein Products

As noted earlier, when the same manufacturing operations are used, the caseins produced from lactic, sulphuric or hydrochloric acid precipitation are almost indistinguishable from one another.

During the acidification process in the manufacture of acid casein, the calcium and inorganic phosphate (that are associated with the casein micelle in milk) are dissolved and leached from the curd leaving only the organic phosphorus and a small residue of calcium. Rennet casein contains about 3% calcium and approximately 1.4% phosphorus.

As they are spray-dried products, their moisture content is much lower than that of the caseins, and their protein content is consequently higher. With a pH generally in the range 6.5-7.0, sodium caseinate will usually contain 1.2-1.4% sodium, whereas the calcium content of calcium caseinate is generally in the range 1.3-1.6%.

Properties of Casein Products

Solubility

Acid and rennet casein are insoluble in water. Virtually all applications of casein products require them to be dissolved first. Consequently, before use, acid casein must be dissolved using an alkali to produce a solution with a pH of 6.5 or higher. The caseinates mentioned in the previous sections are used for food and pharmaceutical applications. For non-food, technical applications, acid casein may be dissolved in other alkalis such as borax or ammonia, usually to a somewhat higher pH (7.5-9.5, or higher) than that used for edible applications.

Water Absorption and Viscosity

Casein products can absorb substantial amounts of water, so they can modify the texture of dough or baked products, serve as the matrix former in cheese-type products, produce specialised plastic materials, or increase the consistency of solutions such as soups. They are good film-formers and find use in whipping and foaming applications, and in emulsions of fats or oils in water.

Nutrition

The nutritional quality of a protein is determined primarily by its essential amino acid content. For adult man, eight amino acids are essential, *i.e.* they must be supplied in the diet.

These are isoleucine, leucine, lysine, methionine, phenylalanine, threonine, tryptophan and valine; the infant requires histidine as well. In comparison with an .ideal. reference protein composition that was developed by the FAO in 1973, casein contains an adequate supply of all the essential amino acids with the possible exception of the sulphur-containing amino acids methionine and cysteine.

Curd

Curds are a dairy product obtained by *curdling* (coagulating) milk with rennet or an edible acidic substance such as lemon juice or vinegar, and then draining off the liquid portion (called whey). Milk that has been left to sour (raw milk alone or pasteurized milk with added lactic acid bacteria or yeast) will also naturally produce curds, and sour milk cheese is produced this way. The increased acidity causes the milk proteins (casein) to tangle into solid masses, or *curds*. The rest, which contains only whey proteins, is the whey. In cow's milk, 80% of the proteins are caseins.

Curd products vary by region and include cottage cheese, quark (both curdled by bacteria and sometimes also rennet) and paneer (curdled with lemon juice). The word can also refer to a non-dairy substance of similar appearance or consistency, though in these cases a modifier or the word *curdled* is generally used (e.g., bean curds, lemon curd, or curdled eggs).

In England, curds produced from the use of rennet is referred to as junket, with true curds and whey only occurring from the natural separation of milk due to its environment (temperature, acidity).

In Asia, curds are essentially a vegetarian preparation using yeast to ferment the milk. In some places in Indian subcontinent, particularly in North India, buffalo milk is used for curd due to its higher fat content, making a thicker curd. The quality of curds depends on the starter used. The time taken to curdle also varies with the seasons, taking less than 6 hours in hot weather and up to 16 hours in cold weather. In the industry, an optimal temperature of 43 °C for 4–6 hours is used for preparation.

In India, the word *curd* means only plain yogurt. In South India, it is common practice to finish any meal with yogurt or buttermilk.

In Sweden, curds is a major ingredient of the traditional baked cheesecakes *Ostkaka* (with egg, flour, almonds), which is eaten with jam and cream.

In Turkey, curds call as ke° and very common in breakfast onto fried bread and also is eaten into macaroni in the region of Bolu and Zonguldak provinces.

Cheese curds, drained of the whey and served without further processing or aging, are popular in some French-speaking regions of Canada, such as Quebec and parts of Ontario. The image to the right shows freshly made morsels of Cheddar cheese before being pressed and aged. In Quebec and Eastern Ontario, cheese curds are popularly served with french fries and gravy as *poutine*. In some parts of the U.S., they are breaded and fried, or are eaten straight. Cheese curds may also be coated with a powdered flavoring agent and sold as a snack food in a fashion similar to flavoured potato chips.

Yoghurt

Yoghurt, yogurt or yogourt (Turkish: *Yoðurt*) is a dairy product produced by bacterial fermentation of milk. The bacteria used to make yoghurt are known as "yoghurt cultures". Fermentation of lactose by these bacteria produces lactic acid, which acts on milk protein to give yoghurt its texture and its characteristic tang.

Worldwide, cow milk is most commonly used to make yoghurt, but milk from water buffalo, goats, sheep, camels and yaks is also used in various different parts of the world. In theory the milk of any mammal could be used to make yoghurt. Soya yoghurt, a non-dairy yoghurt alternative, is made from soy milk; this is not an animal product, being made from soy beans.

Dairy yoghurt is produced using a culture of *Lactobacillus delbrueckii* subsp. *bulgaricus* and *Streptococcus salivarius* subsp. *thermophilus* bacteria. In addition, *Lactobacillus acidophilus*, *Lactobacillus bifidus* and *Lactobacillus casei* are also sometimes used in culturing yoghurt.

The milk is first heated to about 80 °C to kill any undesirable bacteria and to denature the milk proteins so that they set together rather than form curds. The milk is then cooled to about 45 °C. The bacteria culture is added, and the temperature is maintained for 4 to 7 hours to allow fermentation.

Etymology and Spelling

The word is derived from Turkish: *yoðurt*, and is related to the obsolete verb *yoðmak* 'to be curdled or coagulated; to thicken'. The letter ð was traditionally rendered as "gh" in transliterations of Turkish. In older Turkish, the letter denoted a voiced velar fricative /c/, but this sound is elided between back vowels in modern Turkish, in which the word is pronounced [joÈu~t]. Some eastern dialects retain the consonant in this position, and Turks in the Balkans pronounce the word with a hard /a/.

In English, there are several variations of the spelling of the word. In Australia and New Zealand "yoghurt" prevails. In the United Kingdom "yoghurt" and "yogurt" are both current, "yogurt" being more common on product labels, and "yoghourt" is an uncommon alternative. In the United States, "yogurt"' is the usual spelling and "yoghurt" a minor variant. In Canada, "yogurt" is most common among English speakers, but many brands use "yogourt," since it is an acceptable spelling in both official languages.

Whatever the spelling, the word is usually pronounced with a short *o* in the UK, with a long *o* in North America, Australia and South Africa, and with either a long or short *o* in New Zealand and Ireland.

History

There is evidence of cultured milk products in cultures as far back as 2000 BC. The earliest yoghurt was probably fermented spontaneously, perhaps by wild bacteria residing inside goatskin bags used for transportation. In the early 1800s, men used yogurt to clean their goats and sheep. Many women also used yogurt to wash their bodies and hair. Yogurt was the best known cleaning agent at the time.

The oldest writings mentioning yoghurt are attributed to Pliny the Elder, who remarked that certain nomadic tribes knew how "to thicken the milk into a substance with an agreeable acidity". The use of yoghurt by medieval Turks is recorded in the books *Diwan Lughat al-Turk* by Mahmud Kashgari and *Kutadgu Bilig* by Yusuf Has Hajib written in the 11th century. Both texts mention the word "yoghurt" in different sections and describe its use by nomadic Turks. An early account of a European encounter with yoghurt occurs in French clinical history: Francis I suffered from a severe diarrhoea which no French doctor could cure. His ally Suleiman the Magnificent sent a doctor, who allegedly cured the patient with yoghurt. Being grateful, the French king spread around the information about the food which had cured him.

Until the 1900s, yoghurt was a staple in diets of people in the Russian Empire (and especially Central Asia and the Caucasus), Western Asia, South Eastern Europe/Balkans, Central Europe, and India. Stamen Grigorov (1878–1945), a Bulgarian student of medicine in Geneva, first examined the microflora of the Bulgarian yoghurt. In 1905, he described it as consisting of a spherical and a rod-like lactic acid bacteria. In 1907 the rod-like bacteria was called *Lactobacillus bulgaricus* (now *Lactobacillus delbrueckii subsp. bulgaricus*). The Russian Nobel laureate biologist Ilya Ilyich Mechnikov, from the Institut Pasteur in Paris, was influenced by Grigorov's work and hypothesised that regular consumption of yoghurt was responsible for the unusually long lifespans of Bulgarian peasants. Believing *Lactobacillus* to be essential for good health, Mechnikov worked to popularise yoghurt as a foodstuff throughout Europe.

Isaac Carasso industrialised the production of yoghurt. In 1919, Carasso, who was from Ottoman Salonika, started a small yoghurt business in Barcelona and named the business Danone ("little Daniel") after his son. The brand later expanded to the United States under an Americanised version of the name: Dannon.

Yoghurt with added fruit jam was patented in 1933 by the Radlická Mlékárna dairy in Prague. It was introduced to the United States in 1947, by Dannon.

Yoghurt was first introduced to the United States in the first decade of the twentieth century, influenced by Élie Metchnikoff's *The Prolongation of Life; Optimistic Studies* (1908); it was available in tablet form for those with digestive intolerance and for home culturing. It was popularised by John Harvey Kellogg at the Battle Creek

Sanitarium, where it was used both orally and in enemas, and later by Armenian immigrants Sarkis and Rose Colombosian, who started "Colombo and Sons Creamery" in Andover, Massachusetts in 1929.

Colombo Yoghurt was originally delivered around New England in a horse-drawn wagon inscribed with the Armenian word "madzoon" which was later changed to "yogurt", the Turkish name of the product, as Turkish was the lingua franca between immigrants of the various Near Eastern ethnicities who were the main consumers at that time. Yoghurt's popularity in the United States was enhanced in the 1950s and 1960s, when it was presented as a health food. By the late 20th century yoghurt had become a common American food item and Colombo Yogurt was sold in 1993 to General Mills, which discontinued the brand in 2010.

Nutritional Value and Health Benefits

Yoghurt is nutritionally rich in protein, calcium, riboflavin, vitamin B6 and vitamin B12. It has nutritional benefits beyond those of milk. People who are moderately lactose-intolerant can consume yoghurt without ill effects, because much of the lactose in the milk precursor is converted to lactic acid by the bacterial culture.

Yoghurt containing live cultures is sometimes used in an attempt to prevent antibiotic-associated diarrhea.

A study published in the *International Journal of Obesity* (11 January 2005) also found that the consumption of low-fat yoghurt can promote weight loss, especially due to the calcium in the yoghurt.

Varieties and Presentation

Dadiah, or Dadih, is a traditional West Sumatran yoghurt made from water buffalo milk. It is fermented in bamboo tubes.

Yoghurt is popular in Nepal, where it is served as both an appetiser and dessert. Locally called *dahi*, it is a part of the Nepali culture, used in local festivals, marriage ceremonies, parties, religious occasions, family gatherings, and so on. The most famous type of Nepalese yoghurt is called *juju dhau*, originating from the city of Bhaktapur. In Tibet, yak milk (technically dri milk, as the word yak refers to the male animal) is made into yoghurt (and butter and cheese) and consumed.

Tarator and Cacýk are popular cold soups made from yoghurt, popular during summertime in Albania, Bulgaria, Republic of Macedonia, and Turkey. They are made with ayran, cucumbers, dill, salt, olive oil, and optionally garlic and ground walnuts. Tzatziki, a thick yoghurt-based sauce similar in concoction to tarator, is popular in Greece. Bulgaria typically calls tzatziki “dry tarator”.

Khyar w Laban (cucumber and yogurt salad) is a popular dish in Lebanon. Also, a wide variety of local Lebanese dishes are cooked with yogurt like “Kibbi bi Laban” etc.

Rahmjoghurt, a creamy yoghurt with much higher fat content (10%) than most yoghurts offered in English-speaking countries (*Rahm* is German for “cream”), is available in Germany and other countries.

Cream-top yoghurt is yoghurt made with unhomogenised milk. A layer of cream rises to the top, forming a rich yoghurt cream. Cream-top yoghurt was first made commercially popular in the United States by Brown Cow of Newfield, New York, bucking the trend toward low- and non-fat yoghurts.

Jameed is yoghurt which is salted and dried to preserve it. It is popular in Jordan.

Zabadi is the type of yoghurt made in Egypt, usually from the milk of the Egyptian water buffalo. It is particularly associated with Ramadan fasting, as it is thought to prevent thirst during all-day fasting.

Raita is a yoghurt-based South Asian/Indian condiment, used as a side dish. The yoghurt is seasoned with cilantro (coriander), cumin, mint, cayenne pepper, and other herbs and spices. Vegetables such as cucumber and onions are mixed in, and the mixture is served chilled. Raita has a cooling effect on the palate which makes it a good foil for spicy Indian dishes.

Dudh is a Sindhi-curd, popular in India. People drink dudh along with food at intervals, to help digestion and make food more delicious. In some places dudh is also served with plain rice.

Dahi is a yoghurt of the Indian subcontinent, known for its characteristic taste and consistency. The word *dahi* seems to be derived from the Sanskrit word *dadhi*, one of the five elixirs, or panchamrita, often used in Hindu ritual. *Dahi* also holds cultural symbolism in many homes in the *Mithilanchal* region of Bihar. It is found in different flavours, two of which are famous: sour yoghurt (*tauk doi*) and sweet yoghurt (*meesti* or *podi doi*). In India, it is often used in cosmetics mixed with turmeric and honey. Sour yoghurt is also used as a hair conditioner by women in many parts of India. *Dahi* is also known as *Thayiru* (Malayalam), *doi* (Assamese, Bengali), *dohi* (Oriya), *perugu* (Telugu), *Mosaru* (Kannada), *Thayir* (Tamil), or *QÙzana a pÙÙner* (Pashto).

Srikhand, a popular dessert in India, is made from drained yoghurt, saffron, cardamom, nutmeg and sugar and sometimes fruits such as mango or pineapple.

Sweetened and Flavoured Yoghurt

To offset its natural sourness, yoghurt can be sold sweetened, flavoured or in containers with fruit or fruit jam on the bottom. If the fruit has been stirred into the yoghurt before purchase, it is commonly referred to in the United States as Swiss-style. Most yoghurts in North America have added pectin, found naturally in fruit, and/or gelatin to artificially create thickness and creaminess at lower cost. This type of adulterated product is also marketed under the name Swiss-style, although it is unrelated to the way yoghurt is eaten in Switzerland. Some yoghurts, often called "cream line," are made with whole milk which has not been homogenised so the cream rises to the top. Fruit jam is used instead of raw fruit pieces in fruit yoghurts to allow storage for weeks.

Sweeteners such as cane sugar or sucralose – for low-calorie yogurts – are often present in large amounts in commercial yoghurt.

In the USA, sweetened, flavoured yoghurt is the most popular type, typically sold in single-serving plastic cups. Typical flavours are vanilla, honey, or fruit such as strawberry, blueberry, blackberry, raspberry, or peach. In recent years, some manufacturers are marketing flavours inspired by desserts, such as chocolate or cheesecake, with many variants. In Australia, flavoured and Greek are the two most popular types of yoghurt, and are usually sold in one-litre tubs.

Strained Yoghurts

Strained yoghurts are types of yoghurt which are strained through a paper or cloth filter, traditionally made of muslin, to remove the whey, giving a much thicker consistency and a distinctive, slightly tangy taste.

Labneh is a strained yoghurt used for sandwiches popular in Arab countries. Olive oil, cucumber slices, olives, and various green herbs may be added. It can be thickened further and rolled into balls, preserved in olive oil, and fermented for a few more weeks. It is sometimes used with onions, meat, and nuts as a stuffing for a variety of pies or kebbeh balls.

Some types of strained yoghurts are boiled in open vats first, so that the liquid content is reduced. The popular East Indian dessert, a variation of traditional dahi called mishti dahi, offers a thicker, more custard-like consistency, and is usually sweeter than western yoghurts.

Strained yoghurt is also enjoyed in Greece and is the main component of *tzadziki* (from Turkish "cacik"), a well-known accompaniment to gyros and souvlaki pita sandwiches.

Beverages

Ayran or dhalla is a yoghurt-based, salty drink popular in Albania, Bulgaria, Turkey, Azerbaijan, Iran, Pakistan, the Republic of Macedonia, Kazakhstan and Kyrgyzstan. It is made by mixing yoghurt with water and (sometimes) salt. The same drink is known as *doogh* in Iran; *tan* in Armenia; *laban ayran* in Syria and Lebanon; *shenina* in Iraq and Jordan; *laban arbil* in Iraq; *majjiga* (Telugu), *majjige* (Kannada), and *moru* (Tamil and Malayalam) in South India; *lassi* in Punjab and all over India. A similar drink, doogh, is popular in the Middle East between Lebanon, Iran and Afghanistan; it differs from ayran by the addition of herbs, usually mint, and is sometimes carbonated, commonly with carbonated water.

Lassi (urdu) is a yoghurt-based beverage originally from the Indian subcontinent that is usually slightly salty or sweet. Lassi is a staple of Punjab. In some parts of the subcontinent, the sweet version may be commercially flavoured with rosewater, mango or other fruit juice to create a very different drink. Salty lassi is usually flavoured with ground, roasted cumin and red chillies; this salty variation may also use buttermilk, and is interchangeably called *ghol* (Bengal), *mattha* (North India), *majjiga* (Andhra Pradesh), *moru* (Tamil Nadu and Kerala), *Dahi paani* (Odisha), *tak* (Maharashtra), or *chaas* (Gujarat). Lassi is also very widely drunk in Pakistan.

Sweetened yoghurt drinks are the usual form in Europe (including the UK) and the US, containing fruit and added sweeteners. These are typically called "drinking / drinkable yoghurt", such as Yop and BioBest Smoothie. Also available are "yoghurt smoothies", which contain a higher proportion of fruit and are more like smoothies. In Ecuador, yoghurt smoothies flavoured with native fruit are served with pan de yuca as a common type of fast food.

Lassi

Lassi is a popular and traditional Punjabi yogurt-based drink of India and Pakistan. It is made by blending yogurt with water or milk and Indian spices. Traditional lassi (also known as salted lassi, or, simply lassi) is a savory drink sometimes flavoured with ground roasted cumin while sweet lassi on the other hand is blended with sugar or fruits instead of spices. In Dharmic religions, yogurt sweetened with honey is used while performing religious rituals. Less common is lassi served with milk and is topped with a thin layer of clotted cream. Lassis are enjoyed chilled as a hot-weather refreshment, mostly taken with lunch. With a little turmeric powder mixed in, it is also used as a folk remedy for gastroenteritis.

Variations

Traditional Mild Salted Lassi

This form of lassi is more common in villages of Punjab & Porbandar, Gujarat (India). It is prepared by blending yogurt with water and adding salt and other spices to taste. The resulting beverage is known as salted lassi.

Sweet Lassi

Sweet lassi is a form of lassi flavoured with sugar, rosewater and/or lemon, strawberry or other fruit juices. Saffron lassis, which are particularly rich, are a specialty of Sindh in Pakistan and Jodhpur and Rajasthan in India. *Makkhaniya lassi* is simply lassi with lumps of butter in it (*makkhan* is the Punjabi, Urdu, Hindi and Gujarati word for butter). It is usually creamy like a milkshake. Drinking sweet lassi can cause drowsiness, which might help people with sleep disorders.

Mango Lassi

Mango lassi is most commonly found in India and Pakistan though it is gaining popularity worldwide. It is made from yogurt, milk or water and mango pulp. It may be made with or without additional sugar. It is widely available in UK, Malaysia and Singapore, due to the sizable Pakistani/Indian minority, and in many other parts of the world. In various parts of Canada, mango lassi is a cold drink consisting of sweetened kesar mango pulp mixed with yogurt, cream, or ice cream. It is served in a tall glass with a straw, often with ground pistachio nuts sprinkled on top.

Bhang Lassi

Bhang (or bhung) lassi is a special lassi that contains *bhang*, a liquid derivative of cannabis (marijuana), which has effects similar to other eaten forms of marijuana. It is legal in many parts of India and mainly sold during Holi, when pakoras containing bhang are also sometimes eaten. Rajasthan is known to have licensed bhang shops, and in many places one can buy bhang products and drink bhang lassis. However, the term "bhang lassie" is more often a misnomer, as bhang is almost always drunk with thandai, which does not include any curd (joghurt), but either milk, water, crushed ice, sugar, kesar pista flavouring (a ready made thandai syrup), saffron, ground almonds, spices, gound melon and poppy seeds and bhang...etc.

Chaas

Chaas or chaach is a salted drink like lassi; however, chaas contains more water than lassi and has the butterfat removed, so its consistency is not as thick as lassi. Salt and Jeera (cumin seeds) are normally added for taste and sometimes even fresh coriander. Fresh ground ginger & green chillies may also be added as seasoning. Chaas is popular in the Pakistani/Indian Punjab regions of Bhakkar, Jalandhar, and D.I. Khan, and in the Indian states Gujarat and Rajasthan, where it is drunk with the main meal. It is known to aid digestion and is an excellent coolant in the Pakistani and Indian summers. It is called 'majjige' in Kannada, 'majjiga' in Telugu and 'moru' in Tamil and Malayalam.

Ayran

A drink in Turkey is similar to Lassi called Ayran. It is also made with yogurt and water. In Iran and Afghanistan a similar drink is called Doogh.

Tahn

Tahn is the Armenian version of the yogurt drink, consumed cold, with or without food. It is diluted with water and is flavoured with salt.

Cultural References

A 2008 print and television ad campaign for HSBC, written by Jeffree Benet of JWT Hong Kong, tells a tale of a Polish washing machine manufacturer's representative sent to India to discover why their sales are so high there. On arriving, the representative investigates a Lassi parlour, where he is warmly welcomed, and finds several washing machines being used to mix Lassi. The owner tells him he is able to "make ten times as much Lassi as I used to!"

On his *No Reservations* television program, celebrity chef Anthony Bourdain visited a "Govt Authorised" Bhang Shop in Jaisalmer Fort, Rajasthan. The proprietor offered him three varieties of bhang lassi: "normally strong, super duper sexy strong, and full power 24 hour, no toilet, no shower."

Buttermilk

Buttermilk refers to a number of dairy drinks. Originally, buttermilk was the liquid left behind after churning butter out of cream. It also refers to a range of fermented milk drinks, common in warm climates (e.g., Middle East, Pakistan, India, or the Southern United States) where fresh milk would otherwise sour quickly. It is also popular in Scandinavia and the Netherlands, despite the colder climates.

Buttermilk may also refer to a fermented dairy product produced from cow's milk with a characteristically sour taste caused by lactic acid bacteria. This variant is made in one of two ways:*cultured* buttermilk is made by adding lactic acid bacteria (*Streptococcus lactis*) to milk; *Bulgarian buttermilk* is created with a different strain of bacteria called *Lactobacillus bulgaricus*, which creates more tartness.

Whether traditional or cultured, the tartness of buttermilk is due to the presence of acid in the milk. The increased acidity is primarily due to lactic acid, a by-product naturally produced by lactic acid bacteria while fermenting lactose, the primary sugar found in milk.

As lactic acid is produced by the bacteria, the pH of the milk decreases and casein, the primary protein in milk, precipitates causing the curdling or clabbering of milk. This process makes buttermilk thicker than plain milk. While both traditional and cultured buttermilk contain lactic acid, traditional buttermilk tends to be thinner whereas cultured buttermilk is much thicker.

Traditional Buttermilk

Originally, buttermilk was the liquid left over from churning butter from cream. Traditionally, before cream could be skimmed from whole milk, it was left to sit for a period of time to allow the cream and milk to separate. During this time, the milk would be fermented by the naturally occurring lactic acid-producing bacteria in the milk. This facilitates the butter churning process since fat from cream with a lower pH will coalesce more readily than that from fresh cream. The acidic environment also helps prevent potentially harmful microorganisms from growing, increasing shelf-life. However, in establishments that used cream separators, the cream would hardly be acid at all. In the Indian subcontinent, buttermilk is taken to be the liquid leftover after extracting butter from churned yogurt (dahi). Today, this is called *traditional buttermilk*. Traditional buttermilk is still common in many Indo-Pakistani households but rarely found in western countries. In Southern India and most areas of the Punjab, buttermilk with added water, sugar and/or salt, asafoetida, and curry leaves is given at stalls in festival times.

Cultured Buttermilk

Commercially available cultured buttermilk is pasteurized and homogenized (if 1% or 2% fat) milk which has been inoculated with a culture of lactic acid bacteria to simulate the naturally occurring bacteria found in the old-fashioned product. Some dairies add colored flecks of butter to cultured buttermilk to simulate the residual pieces of butter that can be left over from the churning process of traditional buttermilk. Buttermilk solids have increased in importance in the food industry. Such solids are used in ice cream manufacture. Adding specific strains of bacteria to pasteurized milk allows more consistent production.

In the early 1900s, cultured buttermilk was labelled *artificial buttermilk*, to differentiate it from traditional buttermilk, which was known as *natural* or *ordinary buttermilk.*

Acidified buttermilk is a related product that is made by adding a food-grade acid (such as lemon juice) to milk.

Chapter 8

Heat Treatments and Pasteurization

The Purpose of Pasteurization

1. To increase milk safety for the consumer by destroying disease causing microorganisms (pathogens) that may be present in milk.
2. To increase keeping the quality of milk products by destroying spoilage microorganisms and enzymes that contribute to the reduced quality and shelf life of milk.

Pasteurization Conditions

Minimum pasteurization requirements for milk products, and are based on regulations outlined in the Grade A Pasteurized Milk Ordinance (PMO). These conditions were determined to be the minimum processing conditions needed to kill *Coxiella burnetii*, the organism that causes Q fever in humans, which is the most heat resistant pathogen currently recognized in milk. Milk can be pasteurized using processing times and temperatures greater than the required minimums.

Pasteurization can be done as a batch or a continuous process. A vat pasteurizer consists of a temperature-controlled, closed vat. The milk is pumped into the vat, the milk is heated to the appropriate temperature and held at that temperature for the appropriate time and then cooled. The cooled milk is then pumped out of the vat to the rest of the processing line, for example to the bottling station or cheese vat. Batch pasteurization is still used in some smaller processing plants. The most common process used for fluid milk is the continuous process.

The milk is pumped from the raw milk silo to a holding tank that feeds into the continous pasteurization system. The milk continuously flows from the tank through a series of thin plates that heat up the milk to the appropriate temperature. The milk flow system is set up to make sure that the milk stays at the pasteurization temperature for the appropriate time before it flows through the cooling area of the pasteurizer. The cooled milk then flows to the rest of the processing line, for example to the bottling station. There are several options for temperatures and times available for continuous processing of refrigerated fluid milk. Although processing conditions are defined for temperatures above 200°F, they are rarely used because they can impart an undesirable cooked flavour to milk.

History of Pasteurization

The process of heating or boiling milk for health benefits has been recognized since the early 1800s and was used to reduce milkborne illness and mortality in infants in the late 1800s. As society industrialized around the turn of the 20th century, increased milk production and distribution led to outbreaks of milkborne diseases.

Common milkborne illnesses during that time were typhoid fever, scarlet fever, septic sore throat, diptheria, and diarrheal diseases. These illnesses were virtually eliminated with the commercial implementation of pasteurization, in combination with improved management practices on dairy farms. In 1938, milk products were the source of 25% of all food and waterborne illnesses that were traced to sources, but now they account for far less than 1% of all food and waterborne illnesses.

Pasteurization is the process of heating a liquid to below the boiling point to destroy microorganisms. It was developed by Louis Pasteur in 1864 to improve the keeping qualities of wine. Commercial pasteurization of milk began in the late 1800s in Europe and in the early 1900s in the United States. Pasteurization became mandatory for all milk sold within the city of Chicago in 1908, and in 1947 Michigan became the first state to require that all milk for sale within the state be pasteurized. In 1924 the U.S. Public Health Service developed the Standard Milk Ordinance to assist states with voluntary pasteurization programs. The Grade A Pasteurized Milk Ordinance (PMO), as it is now called, is administered by the U.S.

Departments of Health and Human Services and Public Health, and the Food and Drug Administration and defines practices relating to milk parlour and processing plant design, milking practices, milk handling, sanitation, and standards for the pasteurization of Grade A milk products. Each state still regulates milk processing within their own state but dairy products must meet the regulations stated in the PMO for products that will enter interstate commerce.

Fluid Milk Production

Fluid Milk Definitions

Fluid milk is an industry term for milk processed for beverage use. Milk, as defined by the U.S. Code of Federal Regulations (CFR), 21 CFR 131.110, is: "the lacteal secretion, practically free from colostrum, obtained from the complete milking of one or more healthy cows. Milk that is in its final package form for beverage use shall have been pasteurized or ultrapasteurized, and shall contain not less than 8.25% solids and not less than 3.25% milk fat. Milk may have been adjusted by separating part of the milkfat therefrom, or by adding thereto cream, dry whole milk, skim milk, or nonfat dry milk. Milk may be homogenized." Milk solids are the non-water components of milk – protein, lactose, and minerals. Sometimes the combination of protein, lactose and minerals is called the solids not fat content, and when the fat is included it is called total solids content.

Although the CFR states that milk is obtained from cows, the production of milk from other dairy animals in the U.S. (goats, sheep, and water buffalo) also is covered in the Grade A Pasteurized Milk Ordinance (PMO).

Standardization

The fat content of milk varies with species (cow, sheep, goat, water buffalo), animal breed, feed, stage of lactation, and other factors. In order to provide the consumer with a consistent product, most milk in the U.S. is standardized.

To achieve standardization, milk is processed through centrifugal separators to create a skim portion and a cream portion of the milk. Separation produces a skim portion that is less than 0.01% fat and a cream portion that is usually 40% fat, although the desired fat content of the cream portion can be controlled by changing settings on the separator. The cream portion is then added back to the skim portion to yield the desired fat content for the product. Common products are whole milk (3.25% fat), 2% and 1% fat milk, and skim milk (< 0.1% fat).

Pasteurization

The majority of U.S. fluid milk is pasteurized using a high temperature short time (HTST) continuous process of at least 161°F

(71.6°C) for 15 seconds. These conditions provide fresh tasting milk that meets the requirements for consumer safety. Higher heat processes, such as ultrapasteurization or aseptic processing, are used to extend the shelf life of refrigerated products or allow for storage at room temperature, respectively, but may impart a cooked flavour to the milk.

Homogenization

The fat in milk is secreted by the cow in globules of non-uniform size, ranging from 0.20 to 2.0 μm. The non-uniform size of the globules causes them to float, or cream, to the top of the container. Milk that is not homogenized is sometimes referred to as "creamline" milk. Pasteurized milk does not necessarily need to be homogenized. However, homogenized milk should be pasteurized to inactivate native enzymes that deteriorate fat (lipases) and cause rancidity, which results in off-flavours and reduced shelf life in milk.

The purpose of homogenization is to reduce the milk fat globules size to less than 1.0 μm which allows them to stay evenly distributed in milk. Homogenization is a high pressure process that forces milk at a high velocity through a small orifice to break up the globules. The result of homogenization is the creation of many more fat globules of a smaller size. The native milk fat globules are covered in a protein membrane that stabilizes the fat phase in the aqueous (water) phase of milk. Although the milk fat globule membrane is disrupted during the homogenization process, it spontaneously migrates back to the fat globules after homogenization. The new globules created during homogenization are spontaneously coated by proteins in the skim phase from the original milk fat globules.

Vitamin Fortification

Fluid milk is often fortified with vitamin A and vitamin D. The package label must declare when milk is fortified.

Whole milk is considered a good source of vitamin A. Vitamin A is a fat soluble vitamin that is found in the fat phase of milk. The vitamin A content that occurs naturally in 2%, 1% and skim milk is less than in whole milk because of the lower fat levels. Nutritional concerns about consumption of lower fat milk in the late 1970s led to the required fortification of vitamin A in lower fat milks. To achieve the nutritional equivalence of whole milk, lower fat milks should be fortified to 300 IU vitamin A per 8 oz serving. The FDA encourages fortification to a level of 500 IU of vitamin A per 8 oz serving, which is 10 % of the recommended daily allowance (RDA).

Vitamin D is a fat soluble vitamin that occurs naturally in milk but at low levels. Because milk is not considered an important natural source of vitamin D in the diet, vitamin D fortification is voluntary. Fortification of milk with vitamin D began in the U.S. in the 1930s to reduce the incidence of rickets in children. Although rickets is not currently a major concern in the U.S., adequate vitamin D is necessary for human health. Vitamin D helps with calcium absorption, has an important role in bone health, and has a protective effect in cancer. Milk may be fortified with vitamin D to a level of 100 IU per 8 oz serving, which is 25% of the RDA.

Specialty Milk Beverages

The dairy industry has developed specialty fluid milk beverages to meet the diverse nutritional needs of consumers. Lactose-reduced and lactose-free milk, and acidophilus milk were developed for people with lactose intolerance (maldigestion). Lactose-reduced and lactose-free milk are processed, prior to packaging, with the lactase enzyme to separate lactose into its component sugars, glucose and galactose. Acidophilus milk contains *Lactobacillus acidophilus*, a probiotic lactic acid bacterium that is beneficial to human health.

The *Lactobacillus acidophilus* bacteria use lactose for an energy source and reduce the amount of lactose present in milk. They also make the lactase enzyme which assists humans with lactose digestion in the small intestine.

Specialty milk beverages are available that are tailored to specific segments of the population. There are milk beverages with added plant sterols aimed at helping to improve cholesterol levels and others that are fortified with protein and calcium designed for adults. There are carbohydrate-reduced and vitamin fortified milk beverages for people watching their weight. Milk beverages targeted for teen athletes are protein fortified and fat-reduced. Milk beverages designed for children are calcium fortified, fat-reduced and flavoured. The flavoured milks compete with soft drinks for children's attention and come in a wide range of flavours from the traditional chocolate and strawberry to milks flavoured like their favourite candy bar or ice cream.

Yogurt Production

Yogurt Definitions

Yogurt is a fermented milk product that contains the characteristic bacterial cultures *Lactobacillus bulgaricus* and *Streptococcus thermophilus*. All yogurt must contain at least 8.25% solids not fat.

Full fat yogurt must contain not less than 3.25% milk fat, lowfat yogurt not more than 2% milk fat, and nonfat yogurt less than 0.5% milk. The full legal definitions for yogurt, lowfat yogurt and nonfat yogurt are specified in the Standards of Identity listed in the U.S. Code of Federal Regulations (CFR), in sections 21 CFR 131.200, 21 CFR 131.203, and 21 CFR 131.206, respectively.

The two styles of yogurt commonly found in the grocery store are set type yogurt and swiss style yogurt. Set type yogurt is when the yogurt is packaged with the fruit on the bottom of the cup and the yogurt on top. Swiss style yogurt is when the fruit is blended into the yogurt prior to packaging.

Ingredients

The main ingredient in yogurt is milk. The type of milk used depends on the type of yogurt – whole milk for full fat yogurt, lowfat milk for lowfat yogurt, and skim milk for nonfat yogurt. Other dairy ingredients are allowed in yogurt to adjust the composition, such as cream to adjust the fat content, and nonfat dry milk to adjust the solids content. The solids content of yogurt is often adjusted above the 8.25% minimum to provide a better body and texture to the finished yogurt. The CFR contains a list of the permissible dairy ingredients for yogurt.

Stabilizers may also be used in yogurt to improve the body and texture by increasing firmness, preventing separation of the whey (syneresis), and helping to keep the fruit uniformly mixed in the yogurt. Stabilizers used in yogurt are alginates (carageenan), gelatins, gums (locust bean, guar), pectins, and starch.

Sweeteners, flavours and fruit preparations are used in yogurt to provide variety to the consumer. A list of permissible sweeteners for yogurt is found in the CFR.

Bacterial Cultures

The main (starter) cultures in yogurt are *Lactobacillus bulgaricus* and *Streptococcus thermophilus*. The function of the starter cultures is to ferment lactose (milk sugar) to produce lactic acid. The increase in lactic acid decreases pH and causes the milk to clot, or form the soft gel that is characteristic of yogurt. The fermentation of lactose also produces the flavour compounds that are characteristic of yogurt. *Lactobacillus bulgaricus* and *Streptococcus thermophilus* are the only 2 cultures required by law (CFR) to be present in yogurt.

Other bacterial cultures, such as *Lactobacillus acidophilus*, *Lactobacillus subsp. casei*, and Bifido-bacteria may be added to yogurt as probiotic cultures. Probiotic cultures benefit human health by improving lactose digestion, gastrointestinal function, and stimulating the immune system.

Cheese Production

Cheese Definitions

Cheese comes in many varieties. The variety determines the ingredients, processing, and characteristics of the cheese. The composition of many cheeses is defined by Standards of Identity in the U.S. Code of Federal Regulations (CFR).

Cheese can be made using pasteurized or raw milk. Cheese made from raw milk imparts different flavours and texture characteristics to the finished cheese. For some cheese varieties, raw milk is given a mild heat treatment (below pasteurization) prior to cheese making to destroy some of the spoilage organisms and provide better conditions for the cheese cultures. Cheese made from raw milk must be aged for at least 60 days, as defined in the CFR, section 7 CFR 58.439, to reduce the possibility of exposure to disease causing microorganisms (pathogens) that may be present in the milk. For some varieties cheese must be aged longer than 60 days.

Cheese can be broadly categorized as acid or rennet cheese, and natural or process cheeses. Acid cheeses are made by adding acid to the milk to cause the proteins to coagulate. Fresh cheeses, such as cream cheese or queso fresco, are made by direct acidification. Most types of cheese, such as cheddar or Swiss, use rennet (an enzyme) in addition to the starter cultures to coagulate the milk. The term "natural cheese" is an industry term referring to cheese that is made directly from milk. Process cheese is made using natural cheese plus other ingredients that are cooked together to change the textural and/or melting properties and increase shelf life.

Ingredients

The main ingredient in cheese is milk. Cheese is made using cow, goat, sheep, water buffalo or a blend of these milks.

The type of coagulant used depends on the type of cheese desired. For acid cheeses, an acid source such as acetic acid (the acid in vinegar) or gluconodelta-lactone (a mild food acid) is used. For rennet cheeses, calf rennet or, more commonly, a rennet produced through microbial bioprocessing is used. Calcium chloride is sometimes added to the cheese to improve the coagulation properties of the milk. Flavorings may be added depending on the cheese. Some common ingredients include herbs, spices, hot and sweet peppers, horseradish, and port wine.

Bacterial Cultures

Cultures for cheese making are called lactic acid bacteria (LAB) because their primary source of energy is the lactose in milk and their primary metabolic product is lactic acid. There is a wide variety of bacterial cultures available that provide distinct flavour and textural characteristics to cheeses.

Starter cultures are used early in the cheese making process to assist with coagulation by lowering the pH prior to rennet addition. The metabolism of the starter cultures contribute desirable flavour compounds, and help prevent the growth of spoilage organisms and pathogens. Typical starter bacteria include *Lactococcus lactis* subsp. *lactis* or *cremoris*, *Streptococcus salivarius* subsp. *thermophilus*, *Lactobacillus delbruckii* subsp. *bulgaricus*, and *Lactobacillus helveticus*.

Adjunct cultures are used to provide or enhance the characteristic flavours and textures of cheese. Common adjunct cultures added during manufacture include *Lactobacillus casei* and *Lactobacillus plantarum* for flavour in Cheddar cheese, or the use of *Propionibacterium freudenreichii* for eye formation in Swiss. Adjunct cultures can also be used as a smear for washing the outside of the formed cheese, such as the use of *Brevibacterium linens* of gruyere, brick and limburger cheeses.

Yeasts and molds are used in some cheeses to provide the characteristic colours and flavours of some cheese varieties. Torula yeast is used in the smear for the ripening of brick and limberger cheese. Examples of molds include *Penicillium camemberti* in camembert and brie, and *Penicillium roqueforti* in blue cheeses.

Ice Cream Production

Definitions

Ice cream is a frozen blend of a sweetened cream mixture and air, with added flavourings. A wide variety of ingredients are allowed in ice cream, but the minimum amounts of milk fat, milk solids (protein + lactose + minerals), and air are defined by Standards of Identity in the U.S. Code of Federal Regulations (CFR), section 21 CFR 135.110 for ice cream, 21 CFR 135.115 for goat's milk ice cream, and 21 CFR 135.140 for sherbet.

Ice cream must contain at least 10% milk fat, and at least 20% total milk solids, and may contain safe and suitable sweeteners, emulsifiers and stabilizers, and flavouring materials. The finished ice cream must weigh at least 4.5 pounds per gallon and there must be at least 1.6 pounds of total solids (fat + protein + lactose + minerals +

added sugar) per gallon, thus limiting the maximum amount of air (called overrun) that can be incorporated into ice cream. There are well-defined labelling requirements for the types of flavours used (natural and/or artificial) and for the presence of egg yolks in the finished product (ice cream can be called custard or "French" if the content of egg yolks is at least 1.4%). Ice cream may also be labelled as reduced fat (25% less fat than the reference ice cream), light (50% less fat than the reference), lowfat (less than 3 g fat/serving), or nonfat (less than 0.5 g fat/serving).

Ice cream is sold as hard ice cream or soft serve. After the freezing process only a portion of the water is actually in a frozen state. Soft ice cream is served directly from the freezer where only a small amount of the water has been frozen. Hard ice cream is packaged from the freezer and then goes through a hardening process that freezes more of the water in the mix.

Ingredients

Milk fat provides creaminess and richness to ice cream and contributes to its melting characteristics. The minimum fat content is 10% and premium ice creams can contain as much as 16% milk fat. Sources of milk fat include milk, cream, and butter.

The total milk solids component of ice cream includes both the fat and other solids. The other milk solids consists of the protein and lactose in milk and ranges from 9 to 12% in ice cream. The nonfat solids play an important role in the body and texture of ice cream by stabilizing the air that is incorporated during the freezing process. Sources of nonfat solids include milk, cream, condensed milk, evaporated milk, dry milk, and whey.

Sweeteners are used to provide the characteristic sweetness of ice cream. Sweeteners also lower the freezing point of the mix to allow some water to reamin unfrozen at serving temperatures. A lower freezing point makes ice cream easier to scoop and eat, although the addition of too much sugar can make the product too soft. Sweeteners used include sugar (sucrose) and corn syrups.

Stabilizers are proteins or carbohydrates used in ice cream to add viscosity and control ice crystallization. Over time during frozen storage small ice crystals naturally migrate together and form larger ice crystals. Stabilizers help to keep the small crystals isolated and prevent the growth of large crystals, which causes ice cream to be coarse, icy and unpleasant to eat. Stabilizers used include alginates (carageenan), gums (locust bean, guar), and gelatins.

Emulsifiers are used to help keep the milk fat evenly dispersed in the ice cream during freezing and storage. A good distribution of fat helps stabilize the air incorporated into the ice cream and provide a smooth product. Emulsifiers used in ice cream include egg yolks and mono-and diglycerides. A wide range of flavourings are used in ice cream. Flavourings include natural and artificial flavours, fruit, nuts, and bulky inclusions such as chocolate chunks and candies.

Dairy Accounting

To facilitate the recording and accounting of milk supplies each supplier should be given a code number. This code should have two elements:

1. A code for the producer and
2. A code for the dairy cooperative or peasants' association of which the producer is a member; a register of producers and their corresponding codes should be kept at the dairy centre.

The code is recorded in the milk record book, with the weight of milk received, and also on the sample bottle. The supplier should be given a copy of this record each week. He/she should be informed each day of the quality of the milk delivered.

Once the products to be made from the milk have been decided and the prices of the products determined, milk price can be calculated as follows, assuming that butter and cottage cheese are the chosen products:

1. Calculate the value of 1 kg of butterfat from the known price of butter, e.g. EB 10.

   Butter comprises 80% butterfat. Other constituents are regarded as having no commercial value. Therefore, the price of 1 kg of butterfat :

   $= 10 \times 100/80 = \text{EB } 12.5$

2. Calculate the value of 1 litre of skim milk.

   Cottage cheese made from fermented skim milk has a value of EB 1.50/kg. Since there is about an 8-fold concentration of casein in the manufacture of cottage cheese from skim milk, assume an average yield of 1 kg of cottage cheese from 8 litres of skim milk. Therefore, each litre of skim milk has a value of 18 cents.

3. Calculation of the value of milk received: Assume the producer delivers 100 litres of milk containing 4% butterfat.

1. Calculate the weight and value of butterfat received:

   The specific gravity of milk is 1.032 kg/litre. Therefore, the weight of milk received:

   = 100 litres × 1.032 = 103.2 kg

   Weight of fat received can be calculated by multiplying the weight of milk received by the fat content:

   = 103.2 × 0.04 = 4.128 kg

   Value of butterfat purchased from the producer is equal to the weight of butterfat received multiplied by the price per kg of butterfat:

   = 4.128 × EB12.5 = EB51.60

2. Calculate the volume and value of the skim milk. While the actual recovery of skim milk may be greater, in commercial practice it is normally assumed that 80% of the whole milk is recovered as skim milk.

   In this case, we therefore recover 80 litres of skim milk with a value of 18 cents/litre.

   Value of skim milk = 80 × 0.18 = EB 14.40

3. To obtain the total value of the milk received, add the values obtained in 3.1 and 3.2:

   EB 51.60 for butterfat

   EB 14.40 for skim milk

   EB 66.00

Therefore, the average value of 1 litre of milk is 66 cents.

It is important to note that, since the butterfat is the most valuable commercial fraction, milk price will vary in proportion to butterfat content. It is assumed that butterfat content can be estimated. In large dairy plants, milk price is based on the content of the major milk constituents. For small-scale milk processors, this is not normally feasible and payment should be based on fat content.

Production costs and depreciation are deducted proportionally from milk price. Other deductions may also be made when calculating the price paid to the producer for milk.

Milk Analysis

Milk analysis is carried out to determine:

- Freshness
- Adulteration

- Bacterial content, and
- Milk constituents for payment calculation.

Sampling

A representative sample is essential for accurate testing. Milk processors usually pay for milk or cream on the basis of butterfat analysis, and a single butterfat test may be used to determine the butterfat content of thousands of litres of milk or cream. Therefore, an accurate and representative sample must be obtained.

Milk must be mixed thoroughly prior to sampling and analysis to ensure a representative sample. If the volume of milk is small, e.g. from an individual cow, the milk may be poured from one bucket to another and a small sample of milk taken immediately. But if large volumes of milk are handled, the milk or cream must be mixed by stirring. However, it is very difficult to obtain a representative sample of milk or cream when a large volume is dumped into a large container. In such a case the milk must be stirred thoroughly and small samples taken from three or more places in the container. For best results, milk or cream must be sampled when it is at a temperature between 15 and 32°C. If the cream is too cool it will be thick and viscous and will be difficult to sample. Sour milk or cream, in which casein has coagulated, must be sampled frequently. Sampling sour milk follows the same procedure as for fresh milk. If the milk or cream has been standing for a long time and a deposit has formed on the surface and sides of the container, it should be warmed while agitating before a sample is removed. For certain analyses, milk samples can be preserved and stored to await analysis. Samples of milk or cream for butterfat analysis can be preserved using formalin, corrosive sublimate or potassium dichromate. For general analyses, formalin is preferred, because the other two increase the solids content of the milk, influencing total solids determination.

Estimation of Milk PH by Indicator

A rough estimate of pH may be obtained using paper strips impregnated with an indicator. Paper strips treated with bromocresol purple and bromothymol blue are sometimes used on creamery platforms as rejection tests for milk. Bromocresol purple indicator strips change from yellow to purple between pH 5.2 and 6.0, while bromothymol blue indicator papers change from straw yellow to blue-green between pH 6.0 and 6.9.

Electrometric Measurement of PH

Electrometric determination of pH depends on the potential difference set up between two electrodes when they are in contact with a test sample. A reference electrode whose potential is independent of the pH of the solution and an electrode whose potential is proportional to the hydronium ion concentration of the test sample are used. Saturated calomel electrodes are usually used as reference electrodes, and glass electrodes are used to measure pH. Instruments which measure the current produced by the difference in potential between the glass and calomel electrodes are called pH meters.

Preparation of the PH Meter

1. The pH meter should be kept in a dry atmosphere.
2. Before using a new glass electrode, or a glass electrode which has been stored for some time, soak the electrode in N/10 HC1 for about 5 hours.
3. Care should be taken not to scratch glass electrodes against the sides of beakers or other hard surfaces during storage or testing.
4. The level of saturated potassium chloride in the calomel electrode should be checked before making pH measurements.
5. Crystals of potassium chloride should be present in the solution within the electrode.
6. The rubber stopper or cap on the filling arm of the calomel electrode should be removed before making a test.

Standardising and using the pH Meter

1. Rinse the electrodes with distilled water and wipe them gently with tissue or filter paper.
2. Set the temperature; use the control knob of the meter to set the temperature of the buffer used to standardise the meter.
3. Standardise the pH meter against a buffer solution of known pH. Use a buffer solution with a pH as close as possible to that of the test solution.
4. Turn the range selector to the pH range covering the pH of the buffer control until the pointer of the meter reads the pH of the buffer.
5. Set the range switch to zero.
6. Before measuring the pH of the test sample, rinse the electrodes with distilled water and dry them.
7. Set the temperature control knob to the temperature of the sample.
8. Place the test sample in position and allow the electrodes to dip into the solution.

9. Switch the range selector knob to the proper range and read the pH.
10. Rinse the electrodes after use and keep the electrode tips in distilled water between tests.

Always follow the manufacturer's instruction for the particular instrument.

Determination of Milk Acidity

The production of acid in milk is normally termed "souring" and the sour taste of such milk is due to lactic acid. The percentage of acid present in dairy products at any time is a rough indication of the age of the milk and the manner in which it has been handled. As mentioned earlier, fresh milk has an initial acidity due to its buffering capacity.

Apparatus

- White enamelled or porcelain cup
- Stirring rod
- A 10 ml or 17.6 ml pipette
- Burette
- Burette-stand.

Reagents

- One percent alcoholic solution of phenolphthalein
- N/10 or N/9 sodium hydroxide.

I. Using N/10 sodium hydroxide.

Procedure

1. Fill the burette with N/10 NaOH and make sure there are no air bubbles trapped in the lower part.
2. Adjust the level of NaOH in the burette to the top mark – the lowest reading being at the upper end.
3. If milk, skim milk or buttermilk is to be tested, place 18 g in the cup using a 17.6 ml pipette. If cream is to be tested, use a 9 ml pipette (for cream weighing about 1 g/ml).
4. Add 3 to 5 drops of phenolphthalein to the sample in the cup.
5. Note the reading of the NaOH in the burette at the lowest point of the meniscus.
6. Allow the NaOH to flow slowly into the cup containing the sample and stir continuously. When a faint but definite pink colour persists, the end-point has been reached.

7. Take the reading of the burette at the lowest point of the meniscus. Subtract the first reading from the second to determine the number of millilitres of alkali (NaOH) required to neutralise the acid in the sample.

Calculation

Percent lactic acid = ml N/10 alkali × 0.0009 × 100/grams of sample

II. Using N/9 sodium hydroxide: Milk, skim milk and buttermilk.

Apparatus

Same as for I.

Reagents

- 1.6% alcoholic solution of phenolphthalein.
- N/9 sodium hydroxide.

Procedure

1. Put 10 ml of milk in a porcelain dish.
2. Add 0.5 ml of 1.6% solution of phenolphthalein.

Titrate with N/9 sodium hydroxide and follow the same procedures as in I.

Calculation

Percent lactic acid = W/V

Where W = volume of N/9 NaOH required (ml) and

V = volume of milk taken for analysis (10 ml)

III. Using N/9 sodium hydroxide: Cream.

Procedure

1. Put 10 ml of cream in a porcelain dish.
2. Add 10 ml of water with the same pipette.
3. Add 0.5 ml of 1.6% phenolphthalein.
4. Titrate with N/9 NaOH.
5. Calculate as in II.

For determination of acidity of cream serum, the fat percentage of the cream should be known, and the calculation is as follows:

Acidity of serum = (acidity of cream × 100)/100 – % fat

Alcohol Test

The alcohol test, together with the acidity test, is used on fresh milk to indicate whether it will coagulate on processing. Milk that contains more than 0.21 % acid, or calcium and magnesium compounds in greater than normal amounts, will coagulate when alcohol is added.

Apparatus

- Ordinary 6-inch (15 cm) test tubes.
- Test-tube racks or blocks of wood with holes bored to fit the test tubes.

Reagents

The only reagent needed is a 75% alcohol solution. This is usually prepared from 95% alcohol by mixing with distilled water in the proportion of 79 parts of 95% alcohol to 21 parts of distilled water.

Procedure

1. Put equal volumes of milk and 75% alcohol in a test tube.
2. Invert the test tube several times with the thumb held tightly over the open end of the tube.
3. Examine the tube to determine whether the milk has coagulated: if it has, fine particles of curd will be visible.

Clot-on-boiling Test

Acidity decreases the heat stability of milk. The clot-on-boiling test is used to determine whether milk is suitable for processing, as it indicates whether milk is likely to coagulate during processing (usually pasteurisation). It is performed when milk is brought to the processing plant — if the milk fails the test it is rejected. The test measures the same characteristics as the alcohol test but is somewhat more lenient (0.22 to 0.24% acidity, as opposed to 0.21 % for the alcohol test). It has the advantage that no chemicals are needed. However, its disadvantage is that at high altitude milk (and all liquids) boils at lower temperature and therefore the test is even more lenient.

Apparatus

- One boiling water bath (a 600 ml beaker on a heater is adequate).
- Test tubes.
- Timer (a watch or clock is adequate).

Reagents

None

Procedure

1. Place about 5 ml of milk in a test tube (the exact amount is not critical), and place the test tube in boiling water for 5 minutes.

2. Carefully remove the test tube and examine for precipitate. The milk is failed if any curd forms.

Butterfat Determination

The main tests used to determine the fat content of milk and milk products are the Gerber and Babcock tests. Automated methods for testing milk are now used in central laboratories and at large processing centres.

The Gerber Test

The procedures outlined below are used to determine the butterfat content of milk, skim milk, buttermilk, cream and whey.

Milk

Apparatus

The apparatus required for butterfat content analysis comprises:

1. Gerber butyrometer calibrated to read 0–8% or 0–5% and graduated at 0.1 % intervals.
2. Butyrometer stoppers.
3. Milk pipette — volume to match the butyrometer in use.
4. 10 ml double-bulb pipette* for pipetting sulphuric acid.
5. 1 ml bulb pipette* for pipetting amyl alcohol.
6. Thermometer to read 1–100°C
7. Water bath.
8. Gerber centrifuge.

*Alternatively, automatic dispensers can be used for delivering 10 ml of sulphuric acid and 1 ml of amyl alcohol.

Reagents

- 1.825 specific gravity sulphuric acid
- Amyl alcohol.

Procedure

1. Mix the milk sample (temperature about 20°C) thoroughly, taking care to minimise incorporation of air. Allow the sample to stand for a few minutes to discharge any air bubbles. Mix gently again before pipetting.
2. Pipette or dispense 10 ml of sulphuric acid into the butyrometer.
3. Pipette the required volume of milk into the butyrometer. Care must be taken to avoid charring of the milk, by ensuring that the milk flows gently down the inside of the butyrometer. It then rests on top of the acid.

4. Pipette or dispense 1 ml of amyl alcohol.
5. Clean the neck of the butyrometer with a tissue or dry cloth.
6. Stopper the butyrometer tightly using a clean, dry stopper. Shake and invert the butyrometer several times until all the milk has been absorbed by the acid.
7. Then place the butyrometer in a water bath at 65°C for 5 minutes.
8. Centrifuge for 4 minutes at 1100 rpm.
9. Return the butyrometer to the water bath for 5 minutes, ensuring that the water level is high enough to heat the fat column.
10. Read the fat percentage. If necessary, the fat column can be adjusted by regulating the position of the stopper.

Hazards

- Sulphuric acid is toxic, highly corrosive and will cause severe burning if it comes in contact with the skin or eyes.
- When mixing the butyrometer contents, considerable heat is generated.
- If the stopper is slightly loose, leakage may occur during mixing, centrifuging or holding in the water bath.

Precautions

- Wear protective eye goggles
- Avoid all spillage and dropping of sulphuric acid from acid dispensers.
- When mixing, hold the butyrometer stopper firmly to ensure that it cannot slip. Use a cloth or glove to protect the hands when mixing.
- Do not point the butyrometer at anyone when mixing.

Skim milk, buttermilk and whey

Apparatus

Standard Gerber butyrometers designed for testing skim milk. The rest of the apparatus is the same as that used for whole milk.

Reagents: The same reagents are required as for whole milk.

Procedure

The procedure is the same as for whole milk up to and including the first centrifuging. The butyrometers are then placed in the water bath at 65°C, stoppers down, for 1 to 2 minutes and again centrifuged for 4 to 5 minutes.

Then they are placed in the water bath for 2 to 3 minutes and read. A check reading is made after they are placed in the water bath for 2 to 3 minutes. The readings obtained must be corrected as follows:

Percentage read on the: butyrometer	Correction
<0.10%	Add 0.05%
0.10 to 0.25%	Add 0.02%
>0.25%	No correction required

Cream

Apparatus

The apparatus required for whole milk, except for the butyrometers and the 11 ml pipette, is supplemented by certain additional items for testing cream. The test bottles are standard Gerber cream butyrometers.

Other items include a balance for weighing to 0.001 or 0.005 g; a stand to support the butyrometers on the balance or a stopper weighing funnel, and a wash bottle containing warm (30–40°C) distilled water.

Reagents

The same as for whole milk.

Procedure

1. Mix the sample thoroughly, though cautiously, to avoid frothing. If the sample is very thick, it should be warmed to between 37.8° and 50°C to facilitate mixing.
2. Weigh 5 g of cream into the butyrometer.
3. Add about 6 ml of warm distilled water from the wash bottle.
4. Add 10 ml of sulphuric acid and I ml of amyl alcohol.

The remaining procedures are the same as for whole milk.

Cheese

Fat determination in cheese is carried out in a similar manner to that for milk.

Apparatus

Gerber cheese butyrometer stamped “3 g cheese”. Other apparatus same as for Gerber milk fat analysis.

Reagents

- Distilled water
- Sulphuric acid
- Amyl alcohol.

Procedure

1. Weigh out 3 ± 0.01 g of cheese on a counter-balanced piece of grease-proof paper.
2. Dispense 10 ml sulphuric acid into the butyrometer. Add 3 ml of water carefully so that it rests on the acid.
3. Wrap the 3 g of cheese in the grease-proof paper to form a cylinder that fits into the butyrometer.
4. Add a further 4 to 5 ml of water.
5. Add 1 ml of amyl alcohol.
6. Stopper the butyrometer securely and shake to dissolve the cheese. (It may be difficult to dissolve the cheese. If difficulty is experienced, place the butyrometer in the heated water bath and remove periodically for mixing until the cheese is fully dissolved.) Cheese butyrometers are centrifuged and read as for milk and cream.

Determination of Milk Specific Gravity

Specific gravity is the relation between the mass of a given volume of any substance and that of an equal volume of water at the same temperature. Since 1 ml of water at 4°C weighs 1 g, the mass of any material expressed in g/ml and its specific gravity (both at 4°C) will have the same numerical value. The specific gravity of milk averages 1.032, i.e. at 4°C 1 ml of milk weighs 1.032 g.

Since the mass of a given volume of water at a given temperature is known, the volume of a given mass, or the mass of a given volume of milk, cream, skim milk etc can be calculated from its specific gravity. For example, one litre of water at 4°C has a mass of 1 kg, and since the average specific gravity of milk is 1.032, one litre of average milk will have a mass of 1.032 kg.

Apparatus

- Lactometer – this is a hydrometer (a device for measuring specific gravity) adapted to the normal range of the specific gravity of milk. It is usually calibrated to read in lactometer degrees (L) rather than specific gravity *per se*. The relationship between the two is:

(L/1000) + 1 = specific gravity (sp. gr.)

Thus, if L = 31, specific gravity = 1.031.

- A tall, wide, glass or plastic cylinder.
- A thermometer – the lactometer may have a thermometer incorporated.

Procedure

1. Heat the sample of milk to 40°C and hold for 5 minutes. This is to get all the fat into a liquid state since crystalline fat has a very different density to liquid fat, and fat crystallises or melts slowly. After 5 minutes, cool the milk to 20°C.
2. Mix the milk sample thoroughly but gently. Do not shake vigorously or air bubbles will be incorporated and will affect the result.
3. Place the milk in the cylinder. Fill sufficiently that the milk will overflow when the lactometer is inserted.
4. Holding the lactometer by the tip, lower it gently into the milk. Do not let go until it is almost in equilibrium.
5. Allow the lactometer to float freely until it reaches equilibrium. Then read the lactometer at the top of the meniscus. Immediately, read the temperature of the milk. This should be 20°C. If the temperature of the milk is between 17 and 24°C, the following correction factors are used to determine L:

Temp. (°C)	17	18	19	20	21	22	23	24
Correction	–0.7	–0.5	–0.3	—	+0.3	+0.5	0.8	1.1

e.g. The lactometer reading is 30.5 and the temperature is 23°C.

Corrected lactometer = Lc = 30.5 + 0.8 = 31.3

Calculations

All calculations always use Lc, the corrected lactometer reading. To calculate the specific gravity, divide the corrected lactometer reading by 1000 and add 1.

In our example: Sp. gr. = (31.3)/ 1000 + 1 = 1.0313

Determination of total solids (TS) and solids-not-fat (SNF) in milk

The total solids content of milk is the total amount of material dispersed in the aqueous phase, i.e.

SNF = TS – % fat.

The only accurate way to determine TS is by evaporating the water from an accurately weighed sample. However, TS can be estimated from the corrected lactometer reading. The results are not likely to be very accurate because specific gravity is due to water, material less dense than water (fat) and material more dense than water (SNF). Therefore, milk with high fat and SNF contents could have the same specific gravity as milk with low fat and low SNF contents.

$$TS = (Lc)/\ 4 + (1.22 \times \text{fat}\ \%) + 0.72$$

$$SNF = TS - \text{fat}\ \%$$

$$Or = Lc\ /\ 4 + (0.22 \times \text{fat}\%) + 0.72$$

It should be noted that the relationship between Lc and TS varies from country to country depending on milk composition. The above formulae are called the Richmond formulae and were calculated for Great Britain.

Determination of moisture content of butter

Apparatus

- Aluminium, platinum, nickel or porcelain cup, flat bottomed, about 3 cm in diameter, and not less than 2.5 cm deep, with a spout.
- A glass stirring rod with widened flat end.
- A spoon or steel blade.
- A butter trier.
- Alcohol lamp or other means of heating the sample
- Accurate moisture balance.
- Iron tripod.
- Asbestos-centre wire gauze.

Procedure

1. Weigh 10 g of butter into the cup. Heat the butter over a low flame until it ceases foaming and a light-brown colour appears. When heating the sample, place the container on the asbestos-centre wire gauze on a tripod. This distributes the heat evenly across the bottom of the cup.
2. After the moisture is driven from the butter, allow the sample to cool and reweigh.

Calculations

Percentage moisture content of the butter is calculated as:

Moisture % = (Original weight – final weight)/ Original weight × 100

Factors Affecting Milk Composition

Farmers are paid for market milk by volume, provided the milk meets minimum standards of composition-not less than 3.2 per cent fat and 11.75 per cent total solids, on a weight to weight basis. There is no minimum requirement for levels of solids-not-fat or protein. Manufacturing milk is bought on its yield of fat and protein.

Milk Composition

Milk from Friesian-Holstein cattle typically contains 87.5 per cent water and 12.5 per cent total solids. The ranges in composition of milk solids are: fat, 3.2 to 4.6 per cent; protein, 2.8 to 3.5 per cent; lactose, 4.2 to 4.8 per cent and minerals 0.6 to 0.8 per cent. The composition of milk is influenced by non-nutritional and nutritional factors.

Non-nutritional Factors Affecting Milk Composition

Breeding

Breeding is of considerable importance, since fat and protein levels in the milk are heritable characteristics.

Gains in milk composition made from breeding are permanent and accumulate from year to year. Benefits of sire and cow selection, and of mating decisions made today, will continue to be realized in all future descendants of the herd. In this respect, selection is a very productive means of improving milk composition.

The use of breeding to improve milk composition must be clearly understood, since selection to improve one production trait may lead to a decline in another. Selection based on milk yield will result in an increase in milk, fat and protein yields, but will reduce fat and protein percentages.

Table: *Predicted responses to selection.*

The more +, the greater the response over time to selection criteria.

	Predicted response in-				
Select for:	*Milk*	*Fat %*	*Fat yield*	*Protein %*	*Protein yield*
Milk	+++	-	+++	-	+++
Fat %	-	++	+	++	0
Fat yield	++	+	+++	-	++
Protein %	-	+	-	++	0
Protein yield	++	0	+++	+	+++
Fat and protein yield	+++	+	+++	+	+++

Similarly, selection of sires on protein or fat percentage only will result in a reduction in milk yield with minimal improvement in protein or fat yield. Because manufacturing milk is paid for on the yield of solids, you should select sires on the basis of fat PLUS protein yield. This will result in an increase in milk yield, fat percentage and yield, and protein percentage and yield. Since the stage of lactation affects the percentage and yields of protein and fat, you need detailed herd test data to select cows for breeding or culling purposes.

Stage of Lactation

The composition of milk varies with the stage of lactation. Cows that calve in good condition produce milk with a high fat and protein content during early lactation. The percentages of both fat and protein decline during the first six to eight weeks of lactation, then progressively rise after the cow becomes pregnant to reach their highest levels in late lactation.

Age

Although fat and protein contents decrease with increasing age, these changes are small. Since the age structure of a herd is not readily changed, the age composition of the herd is unlikely to contribute significantly to herd variation in milk composition.

Seasonal Conditions

Environmental factors that affect feed intake can be associated with pronounced variations in milk yield and composition. Temperatures consistently above 30C will reduce milk yield as well as the percentage of milk protein, because of a reduction in energy intake. Cows in early to mid-lactation and receiving little or no supplementation (that is, relying on high pasture intakes) will be affected the most by heat stress.

Mastitis

Clinical and subclinical mastitis decrease milk yield and so reduce fat and protein yields.

Nutritional Factors

Level of Feeding

Precalving: Increased feed intake in late pregnancy increases milk yield and the yields of fat and protein. Research has shown that for each 30 kg increase in liveweight at calving, milk yield increases by 122 kg, fat yield by 8 kg and protein yield by 4 kg during the first 20 weeks of lactation. The effects of condition score at calving on fat and protein percentage are small.

Post-calving: The effect of feeding level on fat and protein percentage is variable. This is because the stage of lactation influences the effect of feed intake on milk composition If feed intake is increased during early lactation, milk yield will increase with consequent increases in fat and protein yields. As intake increases, the percentage of milk fat will decline, but protein percentage will increase slightly. Protein production in well fed herds is rarely below 3.2 per cent, but in poorly fed herds it can fall to 2.8 per cent.

Diet Quality

Pasture species influence milk yield and composition, as shown in table below. The use of species associated with improved pasture quality results in increased milk, fat and protein yields.

Table: *Effect of pasture species on milk production, as shown in two trials*

	Trial I		*Trial 2*	
	Kikuyu	*Ryegrass*	*Ryegrass*	*Clover*
Milk yield (L/day)	13.4	19.4	16.5	18.9
Milk Fat (%)	3.7	3.5	3.7	3.5
Milk Fat kg/day	0.51	0.70	0.59	0.69
Protein %	2.9	3.2	3.0	3.2
Protein kg/day	0.40	0.64	0.51	0.62

The increase in fat yield is caused by an increase in milk yield only, since the percentage of milk fat actually declines. increases in both milk yield and percentage protein cause the increase in protein yield. Species differences are largely caused by inherent differences in intake. However, with ryegrass and clovers, differences still occur when they are fed at the same level.

Concentrates

Providing supplementary feed in the form of cereal grain usually results in increased milk, fat and protein yields. An increase in milk yield causes the increase in fat yield, since the percentage of milk fat often declines.

Increases in both milk yield and protein percentage cause the increase in protein yield.

Feeding lupins also results in increases in milk fat and protein yields. Unlike the cereal grains, lupins do not reduce the fat percentage when they are fed as a supplement to cattle. The cow generally uses protein supplements as a source of energy rather than a supply of protein to the udder. Providing there is sufficient protein in the total diet, feeding protein supplements will result in a similar increase in protein percentage as feeding a similar amount of energy from cereal grain.

If the protein content of the total diet is low, feeding protein supplements increases the energy content of the total diet by increasing the digestibility of the total diet. As a result, the protein percentage of the milk is increased. Milk production and milk protein content will increase when an energy deficiency is corrected.

When concentrates are fed, the degree of processing can affect the fat percentage of the milk. Grains need only be cracked to allow sufficient digestion. Over-processing can reduce the fat percentage of the milk.

Fibre

If the milk fat percentage has dropped, but the protein percentage has remained constant, more fibre is needed in the total diet. This is best provided by feeding hay. However, cattle need only small quantities of hay (2 to 3 kg/cow/day)when they are grazing good quality pasture.

Summary

To increase fat and protein yields by feeding, increase the energy intake of the cow by:

- greater pasture intake, by increasing pasture availability (for example, by using irrigation or more nitrogenous fertilizers);
- greater pasture intake by improving the pasture quality, by species selection (ryegrass/clover) or improved grazing management (short compared with rank pasture); or
- by feeding supplements.

Cow's Genetic Predisposition Affects Composition of Her Milk

The genetic predisposition of cows has an effect on the fat and protein content of their milk. Researchers at Wageningen University have spent the past few years examining the scope and significance of genetic variation between cows for the differences in quality characteristics of milk. They have discovered a number of genes that contribute to this genetic variation.

The research was carried out as part of the large-scale Milk Genomics project that Wageningen University launched in 2004 in association with the cattle breeding and dairy sector. On the basis of this knowledge, it is possible to devise an innovative breeding programme for cows and bulls to increase the proportion of unsaturated fatty acids in the milk and to improve cheese production.

Researchers found enormous variation in the composition of the milk fat in cows' milk. A significant proportion of these differences can be put down to genetic predisposition. DNA was analysed to find out which genes contribute to genetic differences between the animals. Researchers in Wageningen demonstrated that a mutation in a gene with a large influence on the amount of fat in milk, also affects the composition of that milk fat. Moreover, the Wageningen researchers were able to make use of available cattle genome data in their research; late last week more than 300 scientists published on this subject in the journal *Science*.

The information on the cattle genome was used to identify new genes that affect the quality characteristics of milk. They identified six areas on the genome, where genes contributing to the genetic variation in milk fat composition are found. According to the researchers, these findings provide an opportunity to devise an innovative breeding programme that exploits the natural variation within the dairy cattle population to make a targeted selection of cattle that produce milk with a modified fat composition. They predicted that the proportion of unsaturated fatty acids in milk could increase by ten percent in ten years by selection of bulls, in addition to the influence that animal nutrition could have on this proportion.

Proteins

The researchers at Wageningen University also discovered substantial variation in the composition of milk proteins, which mainly comprise caseins. Here too, the differences can largely be put down to genetic variation.

DNA analysis revealed three areas on the cow genome affecting the protein composition. Unlike the composition of milk fat, the effect of feed on the differences in the composition of milk protein is relatively small. A higher proportion of caseins results in increased cheese production, which at a rough estimate represents an extra 25 million Euros for the Dutch dairy sector.

Five years ago, researchers at Wageningen University together with the NZO (Dutch Dairy Association) and CRV, an organisation that focuses on cattle improvement, set up the Milk Genomics initiative with the aim of identifying the genes that contribute to natural variation in the quality aspects of milk, and more specifically in composition of fat and proteins. The database created for the genetic research is unique in the world in terms of the scope and number of data. It consists of data on around 2,000 cows from 400 farms.

Milk Composition

Today, not only the nutritional value of milk but also other physiological properties of milk components have attracted interest. Bovine milk contains approximately 87% water, 4.6% lactose, 3.4% protein, 4.2% fat, 0.8% minerals and 0.1% vitamins. The composition of milk continuously undergoes changes depending on e.g. breeding, feeding strategies, management of the cow, lactation stage and season.

Milk Fat

The lipids in bovine milk are mainly present in globules as an oil-in-water emulsion. These fat droplets are formed by the endoplasmic reticulum in the epithelial cells in the alveoli and coated with a surface material of proteins and polar lipids. When secreted, they are enveloped with the plasma membrane of the cell. Membrane-associated materials can comprise 2–6% of the globule mass. The composition and structure of the milk fat globule membrane (MFGM) is not known in detail but it is mainly composed of polar lipids and membrane-bound and associated proteins. The lipid fraction comprising approximately 30% of the membrane material consists of lipids such as phospholipids (25%), cerebrosides (3%) and cholesterol (2%). The remaining 70% of the membrane material are proteins, many of them being enzymes.

The milk fat consists mainly of triglycerides, approximately 98%, while other milk lipids are diacylglycerol (about 2% of the lipid fraction), cholesterol (less than 0.5%), phospholipids (about 1%) and free fatty acids (FFA) (about 0.1). In addition, there are trace amounts of ether lipids, hydrocarbons, fat-soluble vitamins, flavour compounds and compounds introduced by the feed. The size of the milk fat globule (MFG) increases with increasing fat content in the milk probably because of a limitation in production of MFGM.

The number of MFG in milk is approximately 10^{10} per mL with a total area of 700 cm^2 per mL of milk. The size of the MFG has crucial influence on the stability and technological properties of milk. Milk lipid globules are resistant to pancreatic lipolysis in the small intestine unless they are first exposed to gastric lipolysis.

Origin of Milk Fatty Acids

The milk fatty acids are derived almost equally from two sources, the feed and the microbial activity in the rumen of the cow. The fatty acid synthesising system in the mammary gland of the cow produces fatty acids with even number of carbons of 4–16 carbons in length and accounts for approximately 60 and 45% of the fatty acids on a molar and weight basis, respectively.

This *de novo* synthesis in the mammary gland is of the 4:0–14:0 acids together with about half of the 16:0 from acetate and β-hydroxybutyrate. Acetate and butyric acid are generated in the rumen by fermentation of feed components. The butyric acid is converted to β-hydroxybutyrate during absorption through the rumen epithelium. Bovine fat contains certain fatty acids with odd number of carbons, such as pentadecanoic acid (15:0) and heptadecanoic acid (17:0). These two fatty acids are synthesised by the bacterial flora in the rumen. The remaining 16:0 and the long-chain fatty acids originate from dietary lipids and from lipolysis of adipose tissue triacylglycerols. Medium- and long-chain fatty acids, but mainly 18:0, may be desaturated in the mammary gland to form the corresponding monosaturated acids.

Fatty acids are not randomly esterified at the three positions of the triacylglycerol molecule. The short-chain acids butyric (4:0) and caproic (6:0) are esterified almost entirely at *sn*-3. Medium-chain fatty acids (8:0–14:0) as well as 16:0 are preferentially esterified at positions *sn*-1 and *sn*-2. Stearic acid (18:0) is selectively placed at position *sn*-1, whereas oleic acid (18:1) shows preference for positions *sn*-1 and *sn*-3.

When consumed by humans, milk triacylglycerols are lipolysed by lingual lipases in the mouth and by both lingual and gastric lipase in the stomach. The lipases preferentially hydrolyse *sn*-3 position fatty acids, and therefore selectively releases the shorter acids. The result is that 4:0–10:0 pass through the stomach wall in decreasing quantities as the molecular weight increases, enter the portal vein, and are transported to the liver where they are oxidised. About 25–40% of the triacylglycerols are digested in the stomach.

Fatty Acid Composition

Milk fat triacylglycerols are synthesised from more than 400 different fatty acids, which makes milk fat the most complex of all natural fats. Nearly all of these acids are present in trace quantities and only about 15 acids at the 1% level or higher.

Many factors are associated with the variations in the amount and fatty acid composition of bovine milk lipids (15, 16). They may be of animal origin, i.e. related to genetics (breed and selection), stage of lactation, mastitis and ruminal fermentation, or they may be feed-related factors, i.e. related to fibre and energy intake, dietary fats, and seasonal and regional effects. The gross composition of milk fat in Swedish dairy milk 2001 was 69.4% saturated fatty acids and 30.6% unsaturated fatty acids.

The content of saturated fatty acids is lowest in the summer when the cows are grazing, and highest in the winter due to indoor feeding. The content of the unsaturated fatty acids shows the opposite pattern with the highest amount in the summer.

The saturated fatty acids present in milk accounts for approximately 70% by weight (5). The most important fatty acid from a quantitative viewpoint is palmitic acid (16:0), which accounts for approximately 30% by weight of the total fatty acids. Myristic acid (14:0) and stearic acid (18:0) make up 11 and 12% by weight, respectively. Of the saturated fatty acids, 10.9% are short-chain fatty acids (C4:0–C10:0). The amounts of butyric acid (4:0) and caproic acid (6:0) on a yearly average are 4.4 and 2.4% by weight of the total fatty acids, respectively, in Swedish dairy milk. These amounts are higher when their proportions are expressed as molar percentages, approximately 10 and 5%, respectively.

Approximately 25% of the fatty acids in milk are mono-unsaturated with oleic acid (18:1) accounting for 23.8% by weight of the total fatty acids in Swedish dairy milk.

Poly-unsaturated fatty acids constitute about 2.3% by weight of the total fatty acids and the main poly-unsaturated fatty acids are linoleic acid (18:2) and α-linolenic acid (18:3) accounting for 1.6 and 0.7% by weight of the total fatty acids. The ratio between omega-6 and omega-3 fatty acids in Swedish milk fat was 2.3:1 in 2001. Milk and meat from ruminants can be an important source of omega-3 fatty acids in the human diet, as is the case in France, where animal products account for about 40% of the intake.

Approximately 2.7% of the fatty acids in milk are trans fatty acids with one or more trans-double bonds (18). The main trans 18:1 isomer is vaccenic acid (VA), (18:1, 11t), but trans double bounds in position 4–16 is also observed in low concentrations in milk fat. VA constitutes approximately 2.7% of the total fatty acid content and varies with season.

Milk fat contains also conjugated linoleic acid (CLA), with many different isomers including rumenic acid (RA) (cis-9, trans-11 CLA) which predominates (75–90% of total CLA). The majority of RA in the milk fat is synthesised endogenously, in the mammary gland through the action of mammary Ä-desaturase on VA. Thus both VA and RA are present in milk and dairy products, generally in the ration of about 1:3.

VA has a double role in metabolism because it is both a trans fatty acid and a precursor for 9c, 11t-CLA. A small amount of the CLA originates from biohydrogenation of unsaturated fatty acids by rumen bacteria. Animal studies and new human data have confirmed the bioconversion of VA into CLA (21, 22). Bovine milk, milk products and bovine meat are the main dietary sources of the RA (23). Milk content of 9c, 11t-CLA varies considerably, but in Swedish milk fat it constitutes about 0.4% of the fat fraction.

What is Milk?

Dairy herds consist of cows which produce large volumes of milk. All cows are female; males are called bulls. The most common dairy breed is the Holstein, the black and white cows often seen in pastures. Cows are mammals and like all mammals produce milk for their young. This is the milk we get from cows.

Where is Milk Produced in BC?

Most of the dairy herds are in the Lower Mainland, southeastern Vancouver Island, and north Okanagan-Shuswap area. 70% of BC's milk production is in the Fraser Valley. A few dairy herds, about 20%, are located in the North Okanagan, East Kootenay and Bulkley Valley/ Cariboo/Peace regions and 10% on South-east Vancouver Island.

How Much Milk do we Produce?

About 842 dairy farms produce an annual volume of over 570 million litres a year. The average herd size is 80 cows plus additional replacement calves and heifers. The average cow on test produces 29L of milk a day and is milked for 10 months a year, which equals 8839 litres of milk per year per cow. That's an average of 100 glasses of milk per day, every day of the year. This amount would fill up 53 bathtubs.

How is Milk Produced?

Before any cow produces milk, she must first become a mother. When a dairy cow reaches about 15 months in age she is bred, usually by artificial insemination. After about 9 months she has a calf and produces milk. After a cow has given birth, she can produce milk for the next 10 months. A cow that is being milked can eat up to 40kg of grass, forage, and hay a day and drink up to 170L of water a day, especially on hot days. That's over a bathtub full. A cow's diet is supplemented with feeds such as barley, wheat, soybean and canola meal. These are formulated and fed according to the energy, protein and other nutritional needs of the animal.

Milking machines are used to milk a cow. The cows go into a milking barn, their udders are cleaned and a rubber lined suction cup is attached to the teat. The suction cup simulates the suckling action of a calf nursing. The suction cup is attached to hoses and pipes which collect the milk in a holding tank. The milk is then quickly cooled. Cows are milked twice and sometimes three times a day, usually at the same times each day. All equipment used for milking is thoroughly cleaned and sanitized before and after each use.

Dairy farms are inspected and certified before it can produce milk. This includes: all milking equipment, milking procedures, milking parlour and barn—everywhere the cows go must be kept clean and well maintained.

Dairy farmers use computers to keep track of how much each cow eats, how much milk each cow produces and even to match a particular cow with a particular bull for breeding. They also use them for finding information (internet) and financial accounting.

What does Milk look Like When We Use It?

We drink fresh milk (whole, 2%, 1%, skim and chocolate) and use milk products such as cheese, yogurt, sour cream, whipping cream, cottage cheese, evaporated milk, sweetened condensed milk and skim milk powder.

Among the cheeses BC produces are cheddar, mozzarella, Parmesan, colby, gouda, farmer, edam, monterey jack, feta, quark, cottage cheese and ricotta.

Milk is 89% water and 11% solids. The nutrients, such as calcium, riboflavin, vitamin A and protein are in the solids. Milk, cheese, and yogurt are an easy way for most people to get the amount of dietary calcium recommended by Health Canada.

What Happens after the Milk Leaves the Farm?

Milk is picked up from the farm by a tanker truck, that is certified before it can carry milk, every second day.

The licensed driver takes a sample from each farm to ensure the milk meets quality and safety standards. Before the milk can be unloaded it is tested for antibiotic residues. If residues are found, the entire shipment is destroyed and the farmer responsible receives a heavy fine (thousands of dollars) and also pays for the entire truckload of milk.

To ensure the safety of milk, it is pasteurized. This is the process of heating milk quickly to 72°C and cooling it very rapidly to 4°C. This kills any harmful bacteria that may find its way into milk. Pasteurizing milk helps keep milk fresh longer by destroying spoilage organisms. The milk is also tested by a certified laboratory for temperature, acidity and flavour, before it is accepted. Bacteria, water contamination and somatic cell counts are tests that are also done regularly. The presence of somatic cell counts is an indicator of animal health and quality. Other tests are carried out from time to time to ensure purity of the product.

In days gone by before homogenization, the cream always rose to the top. Today, most milk is homogenized. Homogenization ensures that the cream is thoroughly mixed throughout the product so that it does not separate out. This process doesn't alter any of the nutrients found in milk.

Throughout the entire process from the time the cow is milked until the milk is packaged, milk is never touched by human hands. Milk is natural-nothing is added except vitamin A and D which is required by law. Milk remains one of the purest and safest foods available.

The dairy (processing plant) is also inspected regularly for cleanliness, handling procedures and equipment standards. All milk contact equipment is cleansed and sanitized on a daily basis-failure to do so would result in bacterial spoilage before the code date. Every dairy and their employees who work in the processing area must be licensed.

Milk is packaged within days, usually within 24 hours, of arriving at a dairy plant. Packaged dairy products are also regularly tested by a certified laboratory for composition to ensure the product contains what it claims.

This is also the final check point to ensure the product meets the standards established of bacteria, coliforms, yeasts, moulds and other potential contaminants. Dairy products at retail outlets are subject to random sampling as a further check of their safety, quality and composition.

The majority of milk produced in BC is sold as fluid milk while the rest is manufactured into semi-fluid and solid products such as cheese, ice cream, yogurt and cottage cheese.

What Producing Challenges do Dairy Producers Face?

Dairy farms are close to being self-sustaining units. Farmers grow two-thirds of the food a cow eats and recycle the cow's manure back to the fields where the feed is grown. Manure is very useful to farmers because it adds nutrients and organic matter which help to sustain and build the quality of the soil.

Further challenges facing today's dairy producers include:

Meeting environmental requirements.

Surviving a market that is increasingly competitive on a global scale.

Increasing input costs for such things as feed (grain), equipment and labour, with decreasing revenue.

Dealing with increasing competition for land use (e.g., urban push, increasing land values, etc.).

Who's Involved in Getting the Milk From the Farm To the Table?

- Dairy farm owner, manager and staff (milkers, herdsmen, field personnel)
- Breed associations
- Artificial insemination technicians
- Dairy herd improvement advisors
- Veterinarians
- Milking equipment, farm equipment, building and facility suppliers
- Feed producers and nutritionists
- Dairy processor field representatives
- Government inspectors and advisors
- Government and university researchers
- Milk tank truck drivers
- Milk product deliverers
- Store employees.

Chapter 9

Nutrition and Maintenance of Dairy

Dairy Foods and Weight Maintenance

It has been well documented that the prevalence of overweight and obesity in both adults and children has increased in recent years. Data from the 2005–06 NHANES survey shows that among United States adults, the prevalence of obesity was 34 percent.

This represents more than double the obesity in the past three decades. In 2003–04, an estimated 66 percent of adults were classified as overweight. Data from two NHANES surveys (1976–1980 and 2003–2004) show that the prevalence of overweight in children and adolescents is increasing, with overweight prevalence reported between 14 – 19 percent.

Some research suggests that consuming three servings of milk, cheese or yogurt each day as part of a nutrient-rich, balanced diet may help maintain a healthy weight. A number of observational studies suggest that those who consume greater amounts of dairy foods tend to weigh less than those who consume less dairy. Other studies have not found consistent results for the relationship between dairy or calcium consumption and weight loss. This is an area that requires continued research.

A review of research published 2008 *Nutrition Reviews* found a positive link between high calcium intake and improved body composition and weight maintenance across a range of ages. In this review, researchers looked at data from more than 90 studies including randomized clinical trials, metabolic experiments and observational studies that assessed the relationship between calcium and/or dairy intake and the proportion of fat and lean body mass in the body. The authors concluded that the majority of research suggests high calcium intake, from sources including dairy foods, may affect body composition by some combination of reducing body fat mass while maintaining lean body mass, reducing weight gain, and increasing weight loss on calorie-restricted diets.

A study published in December 2008 Journal of the American College of Nutrition pertaining to adolescents and body composition found that adolescents (ages 12-16 years) with higher daily dairy food consumption had less body fat and a lower body mass index (BMI) than those with lower daily dairy consumption. The goal of this analysis was to explore the association between dairy consumption and the level of body fat among more than 10,000 children and adolescents using two of the National Health and Nutrition Examination Surveys (NHANES), 1988-1994 and 1999-2002.

Many studies have been done to examine the link between dairy intake and weight. A 2004 study by Zemel et al concluded that individuals who consume three servings of dairy each day while on a reduced-calorie diet lose significantly more weight than individuals consuming the same calorie-level low-dairy diets. It is unclear which component of dairy foods is responsible for the effect on body weight.

A study by Lin et al examined the effect of calcium on weight loss in adult women and found that weight loss was associated with calcium specifically from dairy sources rather than supplemental sources. Other studies suggest that adequate calcium intakes decrease the risk of overweight individuals becoming obese or developing metabolic syndrome.

There are several possible explanations for how dairy and calcium affect body weight. There is some evidence that calcium, specifically from dairy products, may help burn fat—which may result in less fat being stored. Some researchers suggest that the protein content of dairy products leads to earlier satiety, which helps with weight maintenance.

Any effect is small, but over time could result in enhanced weight loss or better weight management. Dairy products contribute only 9 percent of the calories in the United States food supply, yet provide a large proportion of several essential nutrients such as protein, calcium, potassium, magnesium and vitamins A, B12 and D. Dairy products (widely available in fat-free or low-fat versions) can play an important role in weight-management plans, while conferring additional health benefits in terms of bone health, hypertension and colon cancer, to name a few.

Nutrition and Dairy Fertility

It's no secret that good reproductive performance on dairies involves a variety of factors, but often components of reproductive programs get blamed when cows don't get pregnant – which may be the wrong thing to focus on.

"People often focus on just the reproductive part of a program at one point in time, such as synchronizing cows," says Pedro Melendez, DVM, PhD, College of Veterinary Medicine, University of Florida. "They focus on GnRH, prostaglandin, timed AI, etc., but aren't looking at what were the situations surrounding parturition and before that, such as did she have difficulty calving, how was the transition and prepartum period management, what was the incidence of retained fetal membranes, hypocalcemia, ketosis, etc. We blame reproductive management, but we often don't look at the other factors that occurred before this time."

Jose Santos, DVM, PhD, Veterinary Medicine Teaching and Research Centre, University of California, believes it is very important that producers, veterinarians and nutritionists evaluate reproductive problems not only by paying attention to the reproductive management but also to other aspects, such as nutrition. "When cows experience excessive metabolic problems after calving, that will influence subsequent fertility and response to the reproductive program implemented in the herd," he says.

Inadequate feeding of transition and early lactation cows that leads to more ketosis, hypocalcemia, displaced abomasum, etc., normally reduces subsequent fertility. In addition, it has been demonstrated that mastitis, which is prevalent early postpartum and can be influenced by nutrition, is negatively associated with conception and pregnancy maintenance in dairy cattle. "Unless these problems are corrected," says Santos, "it is unlikely that the reproductive program will be successful."

Melendez agrees and says studies relating calving-related disorders to fertility found that cows with hypocalcemia or milk fever were six times more likely to develop a retained fetal membrane and probably three times more likely to develop metritis. And indirectly, those cows are more likely to develop ketosis. "Finally, we end up with an animal that is cycling very late, so it's very important to pay attention to the basic concept and look back to see what's going on in the transition period," says Melendez.

Nutrition can influence fertility at all stages of the reproductive cycle, but the period between late gestation and first postpartum insemination is probably the most critical. Santos says this is when a cow experiences most of the diseases during the entire lactation cycle and what happens to her in this period will determine how successful the subsequent breeding will be.

Melendez echoes that point of view. "Starting in the dry-off and transition period, the last 21 days before and after calving are the most important time periods for nutrition. The more deficiencies we have at that time, the more infertility we're going to have in the herd. If we have a dry-off and dry period in bad shape, the transition period won't improve anything."

Fertility Problems and Nutrition

Nutrition can have remarkable influences on fertility. Inadequate energy intake and/or inadequate body fatness early postpartum can retard resumption of ovulatory cycles, which extends the interval from calving to first estrus, reduces conception rates at first postpartum AI and increases late embryonic losses after first postpartum AI.

In addition to energy status, there are some well-characterized effects of nutrition on fertility. Nutritional deficiencies, such as antioxidant nutrients (vitamin E and Se), influence immunity status of the uterus and the ability of white blood cells to kill bacteria that invades the postpartum uterus. "This makes the cow more likely to develop retained fetal membranes and metritis, both of which reduce fertility," explains Santos.

Another important example is nutritional management during the transition period to prevent subclinical and clinical hypocalcemia. Hypocalcemia can be minimized by manipulating the mineral composition of the prepartum diet (feeding low Na and K diets and including acidogenic salts). Hypocalcemia results in increased blood cortisol concentrations after calving, impairs uterine involution, is associated with increased uterine infections and reduces fertility.

On the other hand, excess of nutrients or dietary components containing toxins can have negative effects on fertility. One example is a diet too high in protein. "In order to get milk production of 25,000 lbs. per lactation, some farms formulate diets too high in crude protein, such as 19 to 20 percent," says Melendez. But he adds that high protein levels lead to high levels of urea and ammonia in the blood, and studies have demonstrated that high levels of urea will affect the uterine environment and can introduce toxic effects to gametes and the embryo.

High-producing lactating dairy cows do not require diets with 19-20 percent crude protein to achieve productions above 26,000 pounds, says Santos. "What these cows really need is a diet that is balanced for the energy and protein components and provides adequate amounts of metabolizable protein with an adequate amino acid profile."

However, producers and nutritionists often conclude that cows need very high crude protein diets. "In fact, well-balanced diets that contain 17 to 18 percent crude protein are enough to provide all the amino acid needs of high-producing cows even in early lactation," explains Santos. "The problem is that high-producing cows that consume a lot of feed will also consume a lot of protein, although the diet is presumptively balanced. Even in those cases, urea nitrogen in the blood will be high as a consequence of the high protein intake."

Feeding excessive amounts of gossypol is another example, adds Santos. "We have shown that feeding cottonseed that results in plasma gossypol concentrations above 5 micograms/mL reduces conception rates and increases fetal losses, and this is, in part, mediated by reduced embryonic development."

When pregnancy gets beyond 25 days, the cow is fairly adept at pro-tecting the conceptus from nutrition-related problems, but even still factors, such as excess copper or iodine, toxic plants, mycotoxins or fungal-infected silage, can produce abortions.

Feeding Heifers Versus Cows

There's a vast difference in how heifers utilize and respond to nutritional factors than cows because heifers are still growing, even when calving at 24 months. Because of that, heifers need a diet rich in protein but with moderate energy to avoid getting too fat and risking dystocia when calving.

Additionally, the pre-calving cow differs from the heifer in that because she's not still growing, in the transition period she's more likely to develop hypocalcemia. "Those animals can be fed acidogenic salts to prevent hypocalcemia," says Melendez. "Heifers are less likely to develop hypocalcemia and don't need acidogenic salts. The take-home message is that there should be two different rations – one for heifers and one for cows." Melendez adds that first-lactation heifers should continue to be fed in a separate group until the second calving when they become mature adults. Santos agrees that heifers prior to calving should be fed differently than cows. In the last three weeks prepartum, heifers have lower dry matter intake relative to their body weight than cows do. Because they are still growing, their requirements for metabolizable protein is greater and, as opposed to cows, increasing the concentration of crude protein in close-up rations above 13 percent benefits postpartum performance.

"The key here is to feed diets that provide adequate metabolizable protein according to the expected intake of the animal," says Santos. Furthermore, prepartum heifers should not be fed acidogenic salts to prevent hypocalcemia as cows are because heifers do not experience this problem and those salts reduce feed intake and might impair the secretion/activity of some regulatory hormones, such as insulin, that are important during transition.

"Whenever possible, cows and heifers should be fed separately pre- and postpartum, and prepartum, they should be fed diets that are tailored to their nutrient needs."

Melendez suggests that the older the animal becomes, the more chronic these nutrition problems can become. "If they are not well-fed, they will probably start with some lameness problems, and sometimes it's so acute they are culled in the first lactation. They are more likely to develop hypocalcemia, chronic metritis, etc."

Using Body Condition Scores

Body condition scores (BCS) indicate the energy reserves of the body. Typically, every cow loses BCS after calving. Cows calving in obese conditions or too thin can have decreased fertility later on for different reasons. For example, an obese cow calving at 3.5 to 4 will eat less, develop more retained fetal membranes and ketosis because they are mobilizing fat, which can then impact cyclicity.

Alternatively, a too-thin cow at calving will not have enough reserves to support milk production and reproduction.

Melendez recommends that at dry-off a cow in general should score at a 3 and gain a.25 to.5 during the dry period so she will calve at 3.25 to 3.5. "Dry cows should never lose body condition score in the dry period," he says. In addition, low energy during the postpartum period can affect the first ovulation and expression of heat.

For heifers, preferred scores at calving are a bit lower, such as 3.25, otherwise dystocia problems can occur. Too much fat can affect the birth canal causing more dystocia.

Body condition's affect on fertility has a lot to do with the change in score from one time point to another. Studies have demonstrated that between calving and the first 60-90 days, the cow should not lose more than 1 point of condition. Beyond that, fertility can be affected. "A cow may be at 2.5 and gaining weight or a cow may be at 2.75 and losing weight," says Melendez. "That cow will have lower fertility even though it has a higher BCS. The change in BCS is more important than just the score itself."

When to Body Condition Score

Melendez likes cows to be scored at calving, between 60-90 days post-calving to determine how much condition is lost in the first 100 days, at mid-lactation and dry-off.

At dry-off, she should score at 2.75 to 3 and be able to recover a quarter or half of a point before calving. Below 2.75 makes it difficult for a cow to gain up to 3.25 by calving time.

At calving, heifers and cows should be at 3.25 to 3.75 and not lose more than.75 units on the 1-5 scale, says Santos. Cows with low BCS at calving have reduced potential for milk production in the first four to six weeks post-partum. "Ketosis and the associated diseases have been negatively associated with conception rates at first postpartum AI."

At 60 days postpartum, pre-breeding, they should be at 2.75 to 3. "This is important because overconditioned animals have less appetite during transition, which makes them more susceptible to ketosis and hepatic lipidosis," says Santos. Cows with BCS less than 2.75 at 60 days postpartum are much more likely to have not experienced any ovulation prior to first postpartum AI (anovular or anestrous cows), which normally reduces conception rates and increases the odds for late embyronic loss, adds Santos.

Melendez does not like to see animals below 2.5 at 90 days post-calving. At mid-lactation, she should score at 2.75 to 3. "Losing just one body condition score can affect her fertility," says Melendez.

Most BCS loss after calving occurs in the first 40 to 60 days; scoring immediately prior to calving or at 40 to 60 days postpartum will allow the producer to evaluate how his/her fresh cow program is. "At mid-lactation, the producer can decide if cows should be fed differently or not if he/she wants to manage cows based on BCS," suggests Santos. "At this point, there still is time to avoid overconditioning or to improve low BCS by moving cows to different groups if different rations are fed to the lactating herd. Lastly, the measurements taken at dry-off and immediately prior to calving will allow the producer to evaluate the dry cow program."

For lactating dairy cows, a key factor beyond nutrition that will influence BCS is the interval from calving to conception. "Cows that become pregnant early in the postpartum period rarely will be overconditioned when they go dry," explains Santos. "However, they can be on the thin side. On the other hand, cows that have prolonged lactations because of poor reproduction are more likely to be overconditioned at next calving."

Heat Stress, Nutrition and Fertility

Heat stress is the most important environmental factor influencing fertility of lactating dairy cattle. Once the body temperature of the cow rises above 40°C (~104°F), fertilization and early embryonic development are impaired. "At this point, there is not much that you can do in terms of nutritional management to improve fertility," says Santos. "What you really have to do is to provide adequate cooling to avoid hyperthermia."

University of Florida research demonstrated that cells under thermal stress are more prone to oxidative stress, which harms cell membranes, and indicated that one possible mechanism by which heat stress influences fertility is by causing more cellular damage to the embryo because of oxidative stress. However, it's not been demonstrated that there is any benefit to increasing the supply of antioxidant nutrients (Vitamin E, B-carotene) on conception rates of heat-stressed cows.

Because heat stress negatively influences feed intake, it might be prudent to increase the energy content of the diet to offset reduced feed intake and minimize energy deficit that impairs reproduction, says Santos. "However, one might be cautious because cows under heat stress are more likely to develop lameness, which can be exacerbated by hotter diets. In addition, Melendez's research demonstrated that high urea nitrogen in the milk of cows reduced fertility when associated with heat stress. This indicates that diets that are unbalanced in protein can be more detrimental to fertility during heat stress.

Nutrition and Expression of Estrus

Textbooks used to say estrus lasts 18 hours, but today's cows don't seem to be reading those old books. "Today you can see a cow in estrus for just two hours," says Melendez. "They are completely different today, as is nutrition and the high levels of production we now have." One of the first things affected by poor nutrition is going to be the expression of estrus.

Poor nutrition, especially inadequate energy intake, can lead to extended postpartum anestrus, so diets that cause excessive body weight loss can reduce expression of estrus. "In addition to that, I am not aware of any research that has evaluated the effect of specific nutrients on intensity, expression and rates of estrus detection in lactating dairy cows," says Santos.

"We and others have demonstrated that 10 to 20 percent of the postpartum cows and 20 to 50 percent of the postpartum heifers have not cycled until 60 days postpartum, and these numbers are highly influenced by body condition score. So, if cows and heifers are underfed early postpartum or experience more postpartum problems that lead to reduced BCS, they will have extended period of lack of estrus expression."

Furthermore, recent studies by Milo Wiltbank's group at the University of Wisconsin have demonstrated that intensity and duration of estrus are negatively correlated with level of milk production. This is one reason today's cows are less likely to be seen in estrus. The increase in milk production has changed, to some degree, the biology of the cow and this has made estrus detection more difficult.

Investigating Problems

A dairy has to have good records to begin with in order to know when reproductive problems are abnormal. A major indicator of fertility in the herd is the pregnancy rate, which is conception rate x heat detection. "Conception rate is affected by nutrition and other factors," says Melendez. "When conception rate drops 5 to 6 percent, that's a red flag. Pregnancy rate changes over time because of seasonality, but when you compare summer-to-summer, winter-to-winter, that number is very important and can indicate problems."

When investigating fertility problems on a dairy, Melendez suggests to first look at the dairy's management in general, heat detection, cow comfort and other factors. "When looking at nutritional influences, work with the nutritionist and see the interactions with rations, the levels of energy, protein, NDF, ADF, nutrients, etc., and see if there have been changes, to start to slowly expose the different pieces of the puzzle."

Today, there are many useful diagnostic tools that can point you in the right direction when analysing nutrition problems. One way is to measure urine pH in cows fed acidogenic salts prior to calving to prevent hypocalcemia. In addition, producers can evaluate ketones in milk by using strips to determine if changes in prevalence of subclinical ketosis early postpartum, says Santos. Other tests that can be used that are associated with the nutritional status of the herd are plasma concentrations of nonesterified fatty acids and ketones in late gestation and early lactation, and milk urea nitrogen (MUN) during lactation. Melendez says high MUN indicates too much protein, and studies have indicated too high MUN might impair fertility.

Obviously, keeping good records of occurrence of disease events is important to evaluate if the health of the herd has not deteriorated. "In addition to these tools, producers, veterinarians and nutritionists should do the common things that normally improve production, such as evaluating forage quality," suggests Santos.

Melendez says monitoring transition cows after calving for 20 days in a separate group is one way to head off problems, then after 21 days, they can go to the milking ration. "But we need to monitor those animals for the first 20 days," he says.

In Florida, the university's Food Animal Group is monitoring temperature after calving within the first 10 days to see if cows have a fever, which is a red flag for metritis. Melendez says, "We're monitoring ketone bodies by taking urine or milk samples to see if they are ketotic. If they are, the more likely they are to develop an LDA. It's a fresh cow program, and it's helped a lot to diagnose these problems at this time."

Other nutrition and management factors come into play in the relationship with fertility. Cow comfort, heat abatement, bunk space, ration mixing, etc., need to also be considered, as is quantity and quality of water. Dry matter content of the TMR is very important as well, says Melendez. Veterinarians and nutritionists need to look at the dry matter content during times of wet weather, or when forages and feeds are wet so they can adjust rations to get the proper dry matter intake.

Santos believes it is critical that veterinarians and nutritionists work together. "By working together, we will be able to make more educated decisions and complement each other on our expertise in problem solving," he says.

"The most important thing to remember is that nutrition and fertility is a highly complex relationship," says Melendez. "And everything is related."

Trace Minerals and Fertility

Adequate balances of major and minor trace minerals and vitamins play important roles in health as well as reproductive efficiency. For example, calcium is involved in any physiological and molecular process. Selenium is very important in preventing retained fetal membranes before calving. Zinc, especially in bulls, is important for semen quality. "In general," says Pedro Melendez, DVM, PhD, "every nutrient in the end is related to fertility."

Trace minerals, such as Cu, Se and Zn, influence the immune system of the cow by being part of enzymes that promote cell integrity at the level of membrane or cytoplasm. This is important for proper postpartum uterine health. Also, vitamins such as E and A have complementary effects to some of these trace minerals – an example is vitamin E and selenium working together as scavengers of free radicals at the membrane and cytoplasm, respectively, and they are both important for proper reproductive performance.

"However, feeding these trace minerals and vitamins in excess of the daily requirements is unlikely to improve reproduction," notes Jose Santos, DVM, PhD. "In fact, excess of some of these trace minerals, such as copper, can be detrimental to fertility."

Nutrition and Bull Fertility

Nutrition is important in dairy bull fertility. Overconditioned bulls can be lazy and not want to mount cows. Thin bulls may have poor semen quality.

The problem is that on dairies using bulls, the bulls consume the same diet as the cows. These diets contains more energy and protein than they need and also contain other ingredients that might be detrimental to bull fertility.

"Bulls fed lactating rations normally become overconditioned, and the fat accumulated in the neck of the scrotum can impair thermoregulation of the testes, which is critical to sperm viability," says Jose Santos, DVM, PhD. In beef cattle, young bulls fed an excess of energy after puberty showed reduced sperm reserves in the epididymus and increased testicular degeneration.

Lactating rations can also contain considerable amounts of cottonseed and cotton byproducts, which are the source of gossypol. Most lactating cows will consume diets with 500 to 1,000 ppm of total gossypol, most of it in the free form if from whole cottonseed, leading to intakes of 15 to 30 g of gossypol/cow/day. "These amounts of gossypol are detrimental to bull fertility as they increase testicular degeneration and sperm abnormalities and might reduce libido," says Santos.

One option is to rotate bulls continuously from the lactating group to a group in which the energy and protein content of the ration is more adequate and contains no gossypol. However, this requires more breeding bulls in the herd.

Effects of Dairy Intake on Weight Maintenance

To compare the effects of low versus recommended levels of dairy intake on weight maintenance and body composition subsequent to weight loss.

Design and Methods

Two site (University of Kansas-KU; University of Tennessee-UT), 9 month, randomized trial. Weight loss was baseline to 3 months, weight maintenance was 4 to 9 months. Participants were maintained randomly assigned to low dairy (< 1 dairy serving/d) or recommended dairy (> 3 servings/d) diets for the maintenance phase. Three hundred thirty eight men and women, age: 40.3 ± 7.0 years and BMI: 34.5 ± 3.1, were randomized; Change in weight and body composition (total fat, trunk fat) from 4 to 9 months were the primary outcomes. Blood chemistry, blood pressure, resting metabolism, and respiratory quotient were secondary outcomes. Energy intake, calcium intake, dairy intake, and physical activity were measured as process evaluation.

Results

During weight maintenance, there were no overall significant differences for weight or body composition between the low and recommended dairy groups. A significant site interaction occurred with the low dairy group at KU maintaining weight and body composition and the low dairy group at UT increasing weight and body fat. The recommended dairy group exhibited reductions in plasma 1,25-$(OH)_2$-D while no change was observed in the low dairy group. No other differences were found for blood chemistry, blood pressure or physical activity between low and recommended dairy groups. The recommended dairy group showed significantly greater energy intake and lower respiratory quotient compared to the low dairy group.

Weight maintenance was similar for low and recommended dairy groups. The recommended dairy group exhibited evidence of greater fat oxidation and was able to consume greater energy without greater weight gain compared to the low dairy group. Recommended levels of dairy products may be used during weight maintenance without contributing to weight gain compared to diets low in dairy products.

Impact of Nutrition on Dairy Cattle Reproduction

Energy is the major nutrient required by adult cattle and inadequate energy intake has a detrimental impact on reproductive activity of the female bovine. Cows under negative energy balance have extended periods of anovulation. Postpartum anestrus, as well as infertility, is magnified by losses of body condition during the early postpartum period.

Resumption of ovulatory cycles is associated with energy balance, but seems to be mediated by a rise in plasma IGF-I; which is linked to nutritional status and concentrations of insulin in blood.

Feeding diets that promote higher plasma glucose and insulin may improve the metabolic and endocrine status of cows. Feeding behaviour of dairy cows during the transition period, particularly a decline in feed intake prior to calving, is associated with risk of postpartum uterine disease, such as metritis. Because metritis has a profound negative effect on risk of pregnancy in dairy cows, providing adequate bunk space and environment to maximize feed intake is expected to minimize the risk of uterine diseases and improve fertility.

Addition of supplemental fat to the diet improves energy intake, modulates PGF2á secretion by the uterus, affects ovarian dynamics, enhances luteal function, and improves fertility. More specifically, some fatty acids (FA) might impact fertilization rate and embryo quality in dairy cows. Although gossypol intake seems to not affect lactation performance of dairy cows, it may affect fertility when the resulting plasma gossypol concentrations are excessive.

Selection of dairy cattle for milk yield has linked the endocrine and metabolic controls of nutrient balance and reproductive events so that reproduction in dairy cattle is compromised during periods of nutrient shortage, such as in early lactation.

The energy costs to synthesize and secrete hormones, ovulate a follicle, and sustain an early developing embryo are probably minimal compared to the energy needs for maintenance and lactation. However, the metabolic and endocrine cues associated with negative energy balance (NEB) impair resumption of ovulatory cycles, oocyte and embryo quality, and establishment and maintenance of pregnancy in dairy cattle.

As the demands for milk synthesis increase, reproductive functions may be depressed when no compensatory intake of nutrients is achieved. Numerous recent studies have reported that reproductive performance is compromised by the nutrient demands associated with high levels of production. Milk yield increases at a faster rate in the first 4 to 6 wk after parturition than energy intake, consequently high yielding cows will experience some degree of negative balance of energy and other nutrients during the early postpartum period. When cows experience a period of NEB, the blood concentrations of nonesterified fatty acids (NEFA) increase, at the same time that insulin-like growth factor-I (IGF-I), glucose, and insulin are low.

These shifts in blood metabolites and hormones might compromise ovarian function and fertility. It has also been reported that energy balance and dry matter intake (DMI) might affect plasma concentrations of progesterone (Vasconcelos et al., 2003; Villa-Godoy et al., 1988), which may interfere with follicle development and maintenance of pregnancy.

During the last decades, genetic selection and improved management of herds have dramatically increased milk production of dairy cows, at the same time that fertility has decreased (Butler, 2003).

Selection for higher milk production in dairy cattle has changed endocrine profiles of cows so that blood concentrations of bovine somatotropin and prolactin have increased; whereas insulin has decreased (Bonczeck et al., 1988). These hormonal changes and the increased nutrient demands for production might negatively impact reproduction of dairy cows. However, adequate nutrition and sound management have been shown to offset depression of fertility in herds with average milk production exceeding 12,000 kg/cow/yr.

Several nutritional strategies have been proposed to improve reproduction of dairy cattle with no detrimental effect on lactational performance. Maximizing DMI during the transition period, minimizing the incidence of periparturient problems, adding supplemental fat to diets, and manipulating the FA content of fat sources are expected to benefit reproduction in dairy cattle. However, factors such as high incidence of metabolic diseases early postpartum, poor body condition score (BCS) at first insemination, and excessive gossypol concentrations in plasma are detrimental to fertility of dairy cattle.

Nutrition and Postpartum Uterine Health and Fertility

Epidemiological studies have clearly demonstrated strong relationships between postparturient diseases and subsequent reproductive performance in dairy cattle. Cows diagnosed with clinical hypocalcemia were 3.2 times more likely to experience retained placenta (RP) than cows that did not have clinical hypocalcemia (Curtis et al., 1983).

Whiteford and Sheldon (2005) also found that hypocalcemia was associated with occurrence of uterine disease in lactating dairy cows.

Markusfeld (1985) reported that 80 per cent of cows with ketonuria developed metritis.

A major risk factor for uterine disease is RP. Generally, cows with RP have increased risk of developing metritis compared with cows not experiencing RP. Both metritis and RP double the risk of cows remaining with uterine inflammation at the time of first postpartum insemination (Rutigliano, 2006). In the US, a recent USDA study (NAHMS, 1996) indicated that the incidence of RP in dairy cows was 7.8 ± 0.2 per cent. A 2006 study on 5 dairy farms in Israel observed that RP was diagnosed in 13.1 per cent (9.4 to 18.1 per cent) and 9.2 per cent (3.6 to 13.8 per cent) of multiparous and primiparous cows, respectively (Goshen and Shpiegel, 2006).

In the same study, metritis affected 18.6 per cent (15.2 to 23.5 per cent) and 30 per cent (19.4 to 42.3 per cent) of the multiparous and primiparous cows, respectively. Both RP and metritis can have devastating effects on reproductive efficiency in lactating dairy cows, with reduced conception rates and extended intervals to pregnancy.

In fact, not only does the clinical disease negatively affect fertility of dairy cows; but subclinical endometritis, a disease characterized by increased proportion of neutrophils in uterine cytology without the presence of clinical signs of inflammation of the uterus, has major deleterious effects on conception rates of lactating dairy cows at first postpartum insemination.

A rising story suggests that feed intake and feeding behaviour around parturition might mediate some of the increased risk for uterine diseases in dairy cattle. Hammon et al. (2006) observed that cows developing uterine disease postpartum experienced reduced DMI beginning 1 wk before calving.

Similarly, cows diagnosed with severe metritis after calving were already consuming less dry matter 2 wk prior to calving (Huzzey et al., 2007). In the same study, even cows that subsequently developed mild metritis had reduced DMI 1 wk before calving compared with cows with healthy uteri. The same group (Urton et al., 2005) observed that cows subsequently developing metritis spent significantly less time eating before and after calving than cows that did not develop metritis. These data indicate that suppressed intake of nutrients or alterations in feeding behaviour prior to calving are major risk factors for development of metritis postpartum.

A potential link between nutrient intake and development of uterine diseases may be the immune status of the cow. Kimura et al.

(2002) evaluated neutrophil function in 142 periparturient dairy cows from 2 herds by evaluating chemotaxic and killing activity of those cells. The authors observed that 14.1 per cent of the cows developed RP. Neutrophils isolated from blood of cows with RP had reduced ability to migrate to placental tissue and reduced myeloperoxidase activity, a marker for oxidative burst and killing activity of neutrophils.

Interestingly, the reduced neutrophil function was observed between 1 and 2 wk prior to calving, which suggests that the reduced innate immune function may be part of the cause of RP rather than a consequence of the disease. In fact, cows that developed uterine disease, either clinical metritis or subclinical endometritis, experienced reduced DMI and neutrophil function prior to calving. These data strongly suggest that inadequate nutrient intake before calving might predispose cows to impaired immune function; and, subsequently, increased risk for uterine diseases that negatively affect reproduction.

Because intake of nutrients seems to influence energy status and immune function of dairy cows, both of which seem to be related to risk of uterine diseases; it is prudent to suggest that nutritional and management strategies that optimize nutrient intake around parturition should improve uterine health and subsequent fertility of dairy cows. Perhaps, of equal or greater importance than the diet composition is the environment to which the preparturient cow is subjected. Inadequate cow comfort, competition for space, and hierarchical status can influence the ability of the cow to consume nutrients; which can consequently predispose her to uterine disease.

Resumption of Postpartum Cyclicity

The onset of lactation creates an enormous drain of nutrients in high producing dairy cows; which, in many cases, antagonizes the resumption of ovulatory cycles. During early postpartum, reproduction is deferred in favour of individual survival. Therefore, in the case of the dairy cow, lactation becomes a priority to the detriment of reproductive functions.

During periods of energy restriction, oxidizable fuels consumed in the diet are prioritized toward essential processes such as cell maintenance, circulation, and neural activity (Wade and Jones, 2004). Homeorhetic controls in early lactation assure that body tissue, primarily adipose stores, will be mobilized in support of milk production.

Therefore, the early lactation dairy cow that is unable to consume enough energy-yielding nutrients to meet the needs of production and maintenance, will sustain high yields of milk and milk components at the expense of body tissues. This poses a problem to reproduction, as delayed ovulation has been linked repeatedly with energy status (Butler, 2003). Energy deprivation reduces the frequency of pulses of luteinizing hormone (LH); thereby impairing follicle maturation and ovulation. Furthermore, undernutrition inhibits estrous behaviour by reducing responsiveness of the central nervous system to estradiol by reducing the estrogen receptor α content in the brain.

Generally, the first postpartum ovulation in dairy cattle occurs 10 to 14 d after the nadir of NEB (Butler, 2003). Severe weight and BCS losses caused by inadequate feeding or illnesses are associated with anovulation and anestrus in dairy cattle. In fact, cows with low BCS at 65 d postpartum are more likely to be anovular (Santos et al., 2008); which compromises reproductive performance at first postpartum insemination.

Prolonged postpartum anovulation or anestrus extends the period from calving to first AI and reduces fertility during the first postpartum service (Santos et al., 2008). In fact, anovular cows not only have reduced estrous detection and conception rates, but also have compromised embryo survival (Santos et al., 2004b). on the other hand, an early return to cyclicity is important in regard to early conception.

The timing of the first postpartum ovulation determines and limits the number of estrous cycles occurring prior to the beginning of the insemination period. Typically, in most dairy herds, fewer than 20 per cent of cows should be anovulatory by 60 d postpartum (Santos et al., 2008). Estrous expression, conception rate, and embryo survival improved when cows were cycling prior to an estrous synchronization program for first postpartum insemination (Santos et al., 2004 a,b).

Resumption of ovarian activity in high producing dairy cows is determined by energy status of the animal. Therefore, feeding management that minimizes loss of body condition during the early postpartum period and incidence of metabolic disorders during early lactation should increase the number of cows experiencing a first ovulation during the first 4 to 6 wk postpartum.

Energy and Reproduction

Energy intake appears to have the greatest impact on energy status of lactating dairy cows. Villa-Godoy et al. reported that variation in energy balance in postpartum Holstein cows was influenced most strongly by DMI (r = 0.73) and less by milk yield (r = -0.25). Therefore, differences among cows in the severity of NEB are more related with how much energy they consume than with how much milk they produce. During periods of NEB, blood concentrations of glucose, insulin, and IGF-I are low; as well as the pulse frequency of GnRH and LH. Plasma progesterone concentrations are also affected by the energy balance of dairy cows.

These metabolites and hormones have been shown to affect folliculogenesis, ovulation, and steroid production in vitro and in vivo. The exact mechanism by which energy affects secretion of releasing hormones and gonadotropins is not well defined; but it is clear that lower levels of blood glucose, IGF-I, and insulin may mediate this process.

It has been suggested that NEB influences reproduction of dairy cows by impacting the quality and viability of the oocyte of the ovulatory follicle and the CL resultant of the ovulation of that follicle.

Because there is substantial evidence that metabolic factors can influence early follicular development, it is conceivable that changes in metabolism during periods of NEB could influence preantral follicles destined to ovulate weeks later during the breeding period. To test this hypothesis, Kendrick et al. randomly assigned 20 dairy cows to 1 of 2 treatments formulated so that cows consumed either 3.6 per cent (high energy) or 3.2 per cent (low energy) of their body weight. Follicles were transvaginally aspirated twice weekly and oocytes were graded based upon cumulus density and ooplasm homogeneity.

Cows in better energy balance (high energy) had greater intrafollicular IGF-I and plasma progesterone levels and tended to produce more oocytes graded as good. Therefore, NEB not only delays resumption of ovulatory cycles, but it might also influence the quality of occytes once cows are inseminated.

Nutritional Manipulation to Increase Energy Intake

Nutritional efforts to minimize the extent and duration of NEB may improve reproductive performance. The first and most important factor that affects energy intake in dairy cows is feed availability (Grant and Albright, 1995).

Therefore, dairy cows should have continual access to a high quality, palatable diet to assure maximum DMI. However, DMI is limited during late gestation and early lactation, which can compromise total energy intake and reproductive performance. Several nutritional management strategies have been proposed to increase energy intake during early lactation. Feeding high quality forages, increasing the concentrate:forage ratio, or adding supplemental fat to diets are some of the most common ways to improve energy intake in cows.

A number of studies have demonstrated the importance of insulin as a signal mediating the effects of acute changes in nutrient intake on reproductive parameters in dairy cattle. In early postpartum dairy cattle under NEB, reduced expression of hepatic growth hormone receptor 1A (GHR-1A) is thought to be responsible for the lower concentrations of IGF-I in plasma of cows (Radcliff et al., 2003). Because IGF-I is an important hormonal signal that influences reproductive events such as stimulation of cell mitogenesis, hormonal production, and embryo development, among other functions; increasing concentrations of IGF-I early postpartum are important for early resumption of cyclicity and establishment of pregnancy.

It is interesting to note that insulin mediates the expression of GHR-1A in dairy cows, which results in increased concentrations of IGF-I in plasma. Because IGF-I and insulin are important for reproduction in cattle, feeding diets that promote greater insulin concentrations should benefit fertility.

Gong et al. (2002) fed cows of low-and high-genetic merit isocaloric diets, that differed in the ability to induce high or low insulin concentrations in plasma. The diets that induced high insulin reduced the interval to first postpartum ovulation and increased the proportion of cows ovulating in the first 50 d postpartum.

Fat, Fatty Acids, and Reproduction

Studies evaluating the effects of supplemental fat on reproductive performance of beef cattle are limited. To our knowledge, no controlled trials have been conducted with adequate number of animals to evaluate the potential for fat supplementation to impact establishment and maintenance of pregnancy of beef cows. De Fries et al. observed a tendency (P = 0.09) for increased pregnancy in Brahman cows fed 5.2 per cent fat compared with cows fed 3.7 per cent fat in the diet; however, the number of cows used in this study was limited to only 20/treatment.

Feeding fat to dairy cattle usually improved the risk for pregnancy, although responses have not been consistent. When fat feeding improved production and increased body weight loss, primiparous cows experienced reduced pregnancy risk at first AI (Sklan et al., 1994); although pregnancy to AI was extremely high in the unsuplemented cows. However, Ferguson et al. observed a 2.2 fold increased risk of pregnancy at first AI and all AI in lactating cows fed 0.5 kg/d of fat, which tended (P = 0.08) to enhance the proportion of pregnant cows at the end of the study (93 vs. 86.2 per cent).

In grazing cows, supplementation with 0.35 kg of FA improved the risk of pregnancy after the first postpartum AI; although a similar proportion of cows were pregnant at the end of the study (McNamara et al., 2003). Feeding calcium salts of long chain fatty acids (Ca-LCFA) of palm oil improved pregnancy of dairy cows (Schneider et al., 1988), although the authors did not report statistical significance. on the other hand, others did not observe improvements in fertility of dairy cows supplemented with Ca-LCFA or oilseeds; which might be attributed to increased milk yield and body weight losses.

Because the benefits of feeding fat may originate from specific FA, others have evaluated whether feeding FA differing in the degree of saturation might influence fertility of cows.

The essential FA of the n-6 and n-3 families are available in much smaller supply to ruminants than nonruminants because of microbial biohydrogenation of FA in the rumen, suggesting that their supplementation may benefit reproduction.

Three recent studies explored the role of n-6 and n-3 FA supplementation to lactating dairy cows on risk of pregnancy after the first postpartum AI. When cows were fed 0.75 kg of fat from flaxseed, a source rich in C18:3 n-3, or sunflower seed, a source rich in C18:2 n-6; pregnancy tended ($P = 0.07$) to be greater for cows fed n-3 FA.

However, a similar response was not observed by others when cows were fed flaxseed as the source of n-3 FA. Similarly, feeding n-3 FA from fish oil as Ca-LCFA did not improve risk of pregnancy in high producing, lactating dairy cows when compared with a source rich in saturated FA or with Ca-LCFA of palm oil.

Juchem evaluated the effect of feeding cows pre-and postpartum Ca-LCFA of either mostly saturated and monounsaturated FA or a blend of C18:2n-6 and trans-octadecenoic FA.

He observed that cows fed unsaturated FA had 1.5 times greater risk of pregnancy either at 27 or 41 d after AI compared with cows fed mostly saturated FA. Improvements in pregnancy risk when cows were fed C18:2 n-6 and trans-octadecenoic FA were supported by improved fertilization and embryo quality in non-superovulated lactating dairy cows.

Because n-3 FA can suppress uterine secretion of PGF2á, it is thought that they have the potential to improve embryonic survival in cattle. In 3 of 5 experiments, feeding n-3 FA either as flaxseed rich in C18:3 n-3 or fish oil rich in eicosapentanoic acid (EPA) and docosahexanoic acid (DHA; Silvestre et al., 2008) reduced pregnancy losses in lactating dairy cows after the first postpartum AI. on the other hand, when n-6 FA were fed as Ca-LCFA, pregnancy losses were similar to those observed for cows fed Ca-LCFA of palm oil.

Collectively, these data suggest that feeding fat to dairy cows generally improves fertility and responses are observed when the energy density of the ration increased with fat feeding.

Also, these data suggest that fertility responses to fat feeding is altered according to the type of FA supplemented in the diet. Feeding n-3 FA from oilseeds has improved pregnancy risk in some, but not all studies; however feeding n-3 FA as Ca-LCFA containing fish oils does not seem to influence risk of pregnancy. on the other hand, feeding Ca-LCFA rich in n-6 and trans-octadecenoic FA improved pregnancy in lactating dairy cows. Although feeding n-3 FA has not consistently improved pregnancy risk, it has reduced pregnancy losses in dairy cows.

Reproduction

During the immediate postpartum period, the cow's immune system is challenged severely, and the innate and humoral defence systems are reduced. The incidence of diseases and disorders can be high during this time period and have a negative impact on reproductive performance. For example the *risk* of pregnancy (odds ratio) was reduced if cows had RP or lost one BCS unit. Reduction in adaptive and innate immunity at parturition increases the risk of health disorders such as RP, metritis, and mastitis.

Selenium has long been associated with immunity. Cattle supplemented with Se-yeast had an 18 per cent increase of Se in plasma in comparison to sodium selenite in some studies. Some regions of the US are deficient in Se, particularly the Southeast; whereas other states, such as California, are mostly adequate in Se.

We have conducted an experiment to evaluate a supplemental source of organic selenium on reproductive and immune responses by dairy cows in FL and CA. objectives were to evaluate effects of organic Se on health and reproductive performance of dairy cows. Cows were assigned prepartum at approximately 25 d prior to expected day of calving to 1 of 2 sources of Se, organic Se or inorganic sodium Se (sodium selenite, SS) fed at 0.3 ppm (DM basis) until 80 d posptartum. In both sites, cows followed the same study protocol and health was monitored daily throughout the study. Rectal temperature was recorded each morning for 10 d postpartum. In FL, vaginoscopic evaluation of the reproductive tract was performed at 5 and 10 d postpartum. Cows were evaluated for incidence of RP, metritis, puerperal metritis, subclinical endometritis by uterine cytology, ketosis, displacement of abomasum, and mastitis. Cows had their ovulation synchronized for first postpartum AI.

Plasma Se concentrations increased with days postpartum, but source of Se did not influence Se concentrations in cows in CA. However, in FL, feeding SY improved plasma Se concentrations (0.087 vs 0.069 ±}.004 μg/ml; $P < 0.01$). Incidence of postpartum diseases did not differ between treatments in both sites, but cows fed SY had smaller incidence of purulent vaginal discharge than those fed SS in FL.

Diet altered frequency of multiparous cows detected with > 1 event of fever (rectal temperature > 39.5 °C; SY, 13.3 per cent (25/188) vs SS, 25.5 per cent (46/181; $P < 0.05$); but the SY effect was not observed in primiparous cows, which had a much higher frequency of fever (40.5 per cent). Vaginoscopy discharge scores at 5 and 10 d postpartum were better for the SY group; namely, 47.1 (217/460) vs 35.0 per cent (153/437) clear, 43.4 [200/460] vs 47.8 per cent [209/437]) mucopurulent, and 9.3 (43/460) vs 17.1 per cent (75/437) purulent for SY and SS groups, respectively ($P < 0.05$). Feeding organic Se (SY) improved uterine health and second service PR during summer.

Diet failed to alter first service pregnancy rates in CA and FL, and second service pregnancy rate in CA. However, second service pregnancy rate in FL was greater for cows fed SY than SS [SY, 17 per cent (34/199) vs SS, 11.3 per cent (24/211); $P < 0.05$]. The benefit of SY on second service pregnancy rate is intriguing. We hypothesize that cows of the SY group were better able to reestablish an embryo-trophic environment at second service following either early or late embryonic losses.

Measures of innate and humoral immune responses were unaltered by source of Se in CA, but cows fed SY in FL had improved neutrophil function and serum titers against ovalbumin. our findings indicated that feeding SY improved measures of humoral and cellular immunity, uterine health, and second service pregnancy rate in cows in FL; which is known as a Se deficient state. However, in CA source of Se had no impact on health, measures of immune response, or reproductive performance.

Gossypol and Reproduction

Gossypol was first discovered by Chinese scientists after noticing that no children were born for more than a decade in a village where people cooked food with cottonseed oil. Since then, innumerous reports in the literature have confirmed the anti-fertility effect of gossypol in mammals. Gossypol disrupts cell membrane metabolism, affects glycolysis, influences mitochondrial and energy metabolism in the cell, and increases fragility of cell membranes, such as in red blood cells. In fact, erythrocyte fragility has been one of the indicators of potential gossypol toxicosis.

Risco et al. were one of the first to show that gossypol can be toxic and even kill growing cattle. They fed rations with 200, 400, or 800 mg/kg of free gossypol (FG) to bull calves for 120 d. The diets with 400 and 800 mg/kg of FG were considered to be toxic and could potentially cause the death of growing ruminants. Baby calves have little ability to detoxify gossypol; therefore toxicity can be easily induced by feeding cotton products.

The negative effects of gossypol on fertility of ruminants are clear in males. Studies at University of Florida and Kansas State University have shown that as little as 8 g/d of FG fed to young bulls reduced sperm quality and sexual activity. However, the female ruminant seems to be relatively insensitive to the anti-fertility effect of gossypol because of rumen detoxification; but *in vitro* data indicate some inhibition of embryonic development and ovarian steroidogenesis.

More recently, a series of experiments by our group demonstrated that consumption of up to 40 mg of FG/kg of bodyweight did not influence follicle and luteal development in dairy heifers, but feeding a diet with 40 mg of FG/kg of body weight reduced embryo quality and development *in vivo* and *in vitro*. These effects likely explain the reduced risk of pregnancy in dairy cows with high plasma gossypol concentrations, and compromised embryo survival after transfer. Therefore, it is prudent to feed lactating dairy cows amounts of cottonseed that result in low plasma gossypol concentrations.

Implications

Inadequate intake of nutrients and inadequate body reserves during early lactation are the major factors affecting reproductive performance of dairy cows. Improving energy balance by increasing energy intake through additional non-fibre carbohydrates or supplemental fat in the diet reduces days to first ovulation and improves conception postpartum. Strong evidence suggests that management of cows during the prepartum period affects uterine health. Inadequate intake of nutrients prepartum and altered feeding behaviour increases the risk of metritis in dairy cows.

Supplementation with unsaturated FA of the n-3 and n-6 families usually improves fertility, as long as it does not interfere with rumen microbial metabolism. It is critical that improved methods to protect these unsaturated FA are required if precise calculations of the supply of unsaturated lipids are to be utilized in dairy cattle ration formulation to improve fertility.

Source of Se might influence health and reproduction of dairy cows, but response seems to be dependent upon the background Se concentrations in dietary ingredients. Lastly, although lactating dairy cows can consume substantial amounts of gossypol with no detrimental effects on health and lactation, when plasma gossypol concentrations exceed 5°/ml, embryo development and establishment and maintenance of pregnancy are compromised.

Bibliography

Alec, William: *The Dairy Chemical Industry*, London: Longman Group Limited, 1971.

Ashby, E.: *Technology and the Academics*, London: Macmillan, 1959.

Bohra, Babita: *Dairy Farming in Mountain Areas*, Daya, Delhi, 2006.

Brock, H.: *History of Dairy Chemistry*, New York: Norton, 1992.

Bucciarelli, L.: *Designing Engineers in Dairy* , Cambridge: MIT Press, 1995.

Chhazllani, V K: *Dairy Chemistry and Animal Nutrition*, Manglam Pub, Delhi, 2008.

Cohen, Lizabeth, *Making a New Deal, Industrial Workers in Chicago, 1919-1939* Cambridge University Press, 1991.

De, Sukumar: *Outlines of Dairy Technology*, Oxford University Press, Delhi, 2001.

Devraj, B: *Impact of Dairy On Small and Marginal Farmers*, Prateeksha Publications, Delhi, 2010.

Droop, H. Richmond: *Laboratory Manual of Dairy Analysis,* Biotech, 2004.

Fox, Patrick F.: *Advanced Dairy Chemistry: Proteins*, New York: Elsevier Applied Science, 1992.

Guarti, Luigi, *The Valuation of Firms*, Blackwell Publishing, 1994.

Gupta, Sudhir: *Hand Book of Dairy Formulations, Processes and Milk Processing Industries*, Engineers India Res Inst, Delhi, 2003.

Hornig, Susanna: *A Grain of Truth: The Media, the Public, and Biotechnology*, Lantham, MD: Rowman and Littlefield, 2001.

Jensen, Robert G.: *Handbook of Milk Composition*, San Diego: Academic Press, 1995.

Julia, F.: *Fruits of Warm Climates*, Miami, Julia F. Morton Publisher, 1987.

Kango, Mangala: *Normal Nutrition: Fundamental and Management*, RBSA, Delhi, 2003.

Kapoor, Ajay: *Dairy Science and Technology,* Vishvabharti Pub, Delhi, 2005.

Koli, P. A.: *Dairy Development in India: Challenges Before Co-Operatives*, Shruti Pub, Delhi, 2007.

Law, Barry A.: *Microbiology and Biochemistry of Cheese and Fermented Milk*, London: Blackie Academic & Professional, 1997.

Marth, Elmer H., and James L. Steele: *Applied Dairy Microbiology*, New York: M. Dekker, 1998.

Meyer, M. W.: *Theory of Organizational Structure*, Indianapolis, 1977.

Miller, Gregory D., Judith K. Jarvis, and Lois D. McBean.: *Handbook of Dairy Foods and Nutrition*, Boca Raton, Fla.: CRC Press, 1999.

Mudgal, V. D.; K. K. Singhal and D. D. Sharma: *Advances in Dairy Animal Production,* International Book Distribut, 2003.

Myers, R.: *Basics of Dairy Chemistry*, Atlantic, Delhi, 2007.

Nedderman, R. M.: *A Handbook of Unit Operations*, Academic, London, 1971.

Nisha, Maimun: *Health, Food and Nutrition*, Kalpaz, Delhi, 2006.

Parihar, Pradeep and Leena Parihar: *Dairy Microbiology*, Agrobios, Delhi, 2008.

Pirtle, Thomas Ross: *History of the Dairy Industry,* Chicago: Mojonnier Bros. Company, 1926.

Porter, A.R., J.A. Sims, and C.F. Foreman: *Dairy Cattle in American Agriculture,* Iowa, Iowa State University Press, 1965.

Qystein, V. Sjaastad: *Physiology of Domestic Animals*, International Book Distributing Co., Delhi, 2005.

Rao, M K: *Food and Dairy Microbiology*, Manglam Pub, Delhi, 2007.

Samvel, A. P. V.: *Agri-Business Management*, Satish Serial Pub, Delhi, 2008.

Sharma, Ramakant: *Chemical and Microbiological Analysis of Milk and Milk Products,* International Book Distributing, 2006.

Shukla , Arvind N.: *Textbook of Dairy Chemistry,* Discovery Pub, Delhi 2010.

Singh, Harmeet: *Dairy Farming*, APH, Delhi, 2005.

Thompson, Paul B.: *Food Biotechnology in Ethical Perspective*, Aspen, CO: Aspen Publishers, 1997.

Tyagi, Prasum: *A Textbook of Animal Physiology*, Dominant, Delhi, 2010.

van, M.A. J. S. Boekel: *Dairy Technology: Principles of Milk Properties and Processes,* New York: M. Dekker, 1999.

Walstra, Pieter, and Robert Jenness: *Dairy Chemistry and Physics*, New York: Wiley, 1984.

Weimar, Mark R. and Don P. Blayney: *Landmarks in the U.S. Dairy Industry*, USDA, ERS, Agriculture Information Bulletin Number 694, 1994.

Index

❑❑❑